T0142491

Wireless Networks

Series Editor

Xuemin Sherman Shen
University of Waterloo
Waterloo, ON, Canada

The purpose of Springer's new Wireless Networks book series is to establish the state of the art and set the course for future research and development in wireless communication networks. The scope of this series includes not only all aspects of wireless networks (including cellular networks, WiFi, sensor networks, and vehicular networks), but related areas such as cloud computing and big data. The series serves as a central source of references for wireless networks research and development. It aims to publish thorough and cohesive overviews on specific topics in wireless networks, as well as works that are larger in scope than survey articles and that contain more detailed background information. The series also provides coverage of advanced and timely topics worthy of monographs, contributed volumes, textbooks and handbooks.

More information about this series at http://www.springer.com/series/14180

Tho Le-Ngoc • Atoosa Dalili Shoaei

Learning-Based Reconfigurable Multiple Access Schemes for Virtualized MTC Networks

 Springer

Tho Le-Ngoc
Electrical and Computer Engineering
McGill University
Montreal, QC, Canada

Atoosa Dalili Shoaei
Electrical and Computer Engineering
McGill University
Montreal, QC, Canada

ISSN 2366-1186   ISSN 2366-1445   (electronic)
Wireless Networks
ISBN 978-3-030-60384-7   ISBN 978-3-030-60382-3   (eBook)
https://doi.org/10.1007/978-3-030-60382-3

This Springer imprint is published by the registered company Springer Nature Switzerland AG.
The registered company address is: Gewerbestrasse 11, 6330 Cham, Switzerland

Preface

Machine-type communications are expected to account for the dominant share of the traffic in future wireless networks. While in traditional wireless networks, designed for human-type communications, the focus is on support of large packet sizes in downlink, machine-type communication systems deal with heavy uplink traffic. This is due to the nature of the tasks performed by machine-type communication devices, which is mainly reporting measured data or a detected event. Furthermore, in these networks, using the virtualization framework, the network infrastructure can be shared between different applications for which providing isolation is of high importance. To support these unique characteristics of machine-type communications, proper access schemes need to be developed, which is the focus of this book.

Leveraging traffic characteristics of the devices can improve the performance of resource utilization of machine-type networks. Therefore, first, we present a traffic-aware carrier sense multiple access (CSMA) scheme with deterministic backoff values, in which the access point assigns backoff values to devices by modeling the problem as a Markov decision process. Furthermore, aiming to improve the performance of the networks, consisting of heterogeneous devices in terms of packet arrival probabilities (i.e., including both frequent and sporadic), we propose a reconfigurable access scheme. In the proposed scheme, time is divided into frames, and each frame is split into two segments: an assignment-based segment and a random access segment. At each frame, the proposed scheme dynamically switches from the assignment-based segment to the random access segment taking into account the traffic parameters of the devices. The assignment-based segment is suitable for devices with high packet arrival probabilities, and the random access segment is more efficient for devices with infrequent packet arrivals. To assign devices to the proper segment, the problem is formulated as a complementary geometric programming problem, where the solution can be obtained by applying a computationally affordable algorithm.

In addition, reinforcement learning algorithms are used for scenarios of unknown traffic parameters. More specifically, for these scenarios, Thompson sampling-based algorithms are proposed and regret analysis is conducted for performance evaluation. Furthermore, the performance of the proposed reconfigurable access scheme is investigated in a scenario in which devices use different wireless technologies to access the channel. In particular, we consider the case that all devices transmit over the same unlicensed channel, where some of them are connected to the long-term evolution (LTE) network while others use WiFi technologies to access the channel. Moreover, to further boost the performance of the proposed reconfigurable access scheme and accommodate connectivity of a large number of devices, the proposed reconfigurable access scheme is enhanced with non-orthogonal multiple access (NOMA) technology. In particular, using optimization techniques, the proposed reconfigurable access scheme, at each frame, chooses which devices to transmit their packets using NOMA, which to transmit in the orthogonal assignment-based segment, and what access parameters to be used for the rest of devices competing in the random access segment. Finally, to reduce the signaling overhead, a self-organized TDMA scheme is proposed in which devices reserve time-slots without the need of a central entity.

This book addresses the main challenges that exist in MTC systems, in particular focusing on the problems related to the design of proper multiple access schemes for these systems.

This work was partially supported by the Natural Sciences and Engineering Research Council (NSERC) and the Huawei Technologies Canada.

Montréal, QC, Canada   Tho Le-Ngoc
Montréal, QC, Canada   Atoosa Dalili Shoaei
May 2020

Contents

Acronyms

AGMA	Arithmetic geometric mean approximation
AP	Access point
BS	Base station
CGP	Complementary geometric programming
CMAB	Combinatorial multi-armed bandit
CSMA	Carrier sense multiple access
DEB	Deterministic backoff
DFA	Demand/free assignment
GP	Geometric programming
LTE	Long-term evolution
M2M	Machine to machine
MAC	Medium access control
MDP	Markov decision process
MTC	Machine-type communication
NERA	NOMA-enhanced reconfigurable access
NOMA	Non-orthogonal multiple access
OMA	Orthogonal multiple access
PDR	Packet delivery ratio
QoS	Quality of service
RA	Random access
RCA	Reconfigurable access
SIC	Successive interference cancellation
TDMA	Time division multiple access
TMAB	Thresholding multi-armed bandit
TS	Thompson sampling
U-LTE	Unlicensed LTE

Chapter 1
Introduction

1.1 Machine-type Communications

With the rapid growth of telecommunication technologies, new use cases and applications are emerging in areas such as remote and mobile health care, elderly assistance, public safety, intelligent energy management and smart grids, smart agriculture, intelligent transportation systems, and so on. Many of these applications involve fully-automated communication between devices, where little or no human intervention is required. This type of communications is generally known as machine-type communications (MTC) [1].

The architecture of MTC systems functions as the basis for supporting machine-to-machine (M2M) applications. A typical MTC architecture consists of three domains: device domain, network domain, and application domain as shown in Fig. 1.1. The device domain is constituted of heterogeneous MTC devices including sensors, actuators, meters, radio frequency identification (RFID) tags, cameras, and others, where each group of devices is serving different purposes, such as measuring the temperature or humidity of the environment, detecting a movement, etc. The MTC devices are connected to the base station (BS)/access point (AP) directly or in a multi-hop manner through MTC gateways. In fact, an MTC gateway operates as a proxy between MTC devices and the network domain. As an example, an MTC gateway can run an application that combines MTC traffic and forwards the aggregated traffic to the BS/AP. The application domain consists of MTC servers with which devices exchange information through the connectivity provided by the network domain [2].

© Springer Nature Switzerland AG 2020
T. Le-Ngoc, A. Dalili Shoaei, *Learning-Based Reconfigurable Multiple Access Schemes for Virtualized MTC Networks*, Wireless Networks, https://doi.org/10.1007/978-3-030-60382-3_1

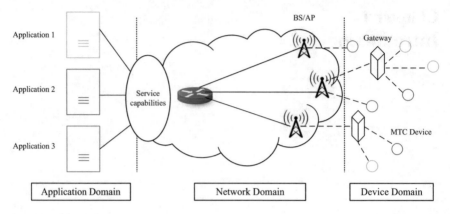

Fig. 1.1 MTC architecture

1.2   Machine-type Applications

Machine-type applications cover a wide variety of domains from industrial automation and control to environmental monitoring towards building an information ambient society [3, 4]. These applications can be categorized as follows.

1.2.1   Industrial Automation and Control

The automation in industrial plants can be enhanced by exchanging and gathering information among sensors, actuators, and RFID tags in M2M communications associated with the products. For example, vibrations in industrial machineries can be monitored by MTC devices and if they exceed a specific threshold, a warning can be issued or the whole production can be made to stop. Upon the occurrence of such an event, the MTC devices transmit the relevant information to the MTC server via the network. Consequently, depending on the event, the MTC controller server will react to the situation [5].

Other scenarios for this category of machine-type applications include production on demand, optimization of packaging, logistics and supply chain, and inventory tracking.

1.2.2   Intelligent Transportation

Another area that MTC can play an important role is transportation. Different types of vehicles including cars, trains, trucks, buses, motor-bikes, and container lorries can be equipped with sensors, actuators, and processing power to become an entity in MTC systems. In addition, roads and transported goods can be equipped with sensors and tags to send information to MTC control servers and transportation companies. This information can be used for traffic routing, monitoring the transported goods status, and tracking the fleets locations. More applications in the transportation domain are discussed below.

1.2.2.1 M2M-Assisted Driving

An example of these types of applications is a driving behavior monitoring system which helps the driver not to fall asleep by generating alerts and warnings. Furthermore, the MTC systems can also automatically call for help if they detect an accident. In addition, using the information of these systems, the road traffic patterns can be obtained for route planning purposes.

1.2.2.2 Fleet Management

MTC technology also brings benefits for fleet management systems in many ways. They enable vehicle tracking to gather the data on locations, fuel consumption, humidity and temperature to increase fleet safety, reduce the accident rates, and increase the productivity of a fleet company. This information helps a fleet business to do a better resource management and more precise control leading to the cost reduction and enabling the business to maintain its competitiveness.

1.2.2.3 e-Ticketing and Passenger Services

In traditional public transportation systems, ticketing is done mainly manually and in some cases semi-automatically. Usually, these systems require labors to do tedious, time-consuming, and stressful tasks. A better option is to use an e-ticketing model. To realize the e-ticketing, the near field communication (NFC) technology can be used in which a mobile phone with NFC capability is scanned to obtain the passenger's identity at the entrance/exit of the station. Each station is identified by a code number. At the station, after scanning the mobile phone, the code number of the station is sent to the server of the transportation service provider through a wireless network. The server computes the fare based on the distance traveled and other relevant metrics and sends the fare to the mobile M2M service provider which deducts the fare amount from the passenger's account. Using such a system can enhance the effectiveness of ticketing, save the costs for transportation service providers, and increase convenience of passenger.

1.2.2.4 Smart Parking

Although car is the most ubiquitous means for humans transportation, it has significantly negatively impacted living conditions. This is due to the long time spent for searching parking spaces which causes air pollution, financial loss and much more. Smart parking is a proven, robust and cost-effective solution to let road users know about the location of unoccupied car parking spaces. In these systems, M2M sensors are employed in parking spaces to detect the cars that are parking over them. More specifically, once a car is detected, the M2M sensor sends that information to the M2M database server via the connected network and the occupancy of the parking space is updated in the corresponding application.

These applications not only make easier to locate the parking spot but also allow drivers to save fuel and associated costs. Furthermore, the city council can use these applications to monitor and manage the parking spaces and obtain real time information.

1.2.3 Smart Grid

Recently, smart grid has been receiving a lot of attention due to its capability to manage electric power consumption. The main feature of smart grid is to use M2M communications to satisfy the needs for enhancing the efficiency of power generation, distribution, and consumption sectors. Furthermore, another goal of using smart grid is to reduce energy wastage to minimize CO_2 emission. To achieve this goal, the energy usage information from various sectors of the smart grid systems should be collected and analyzed in real time to configure the operating parameters to achieve this goal.

1.2.4 Smart Environment

The quality of human life can significantly be improved by implementing automation in every aspect of the daily life. To this end, the M2M communications can be used to collect the information generated at various places including homes, offices, and different corners of cities. This real-time information can help people to make better decisions leading to reduced living costs, and more efficient utilization of natural resources. Some of the scenarios realizing smart environment are discussed below.

1.2.4.1   Smart Homes, Offices, and Shops

Our living environment is surrounded by various electronic appliances including lights, air conditioners, heaters, refrigerators, microwave ovens, and cookers. These appliances are equipped with sensors and actuators to make more efficient utilization of energy and also to make our life more convenient. For example, heating and cooling can be adapted according to the weather conditions to maintain a desirable temperature. Furthermore, the lighting in rooms can be adjusted based on the time of a day and to the number of occupants inside the rooms. In addition, incidents such as fire, a fall of elderly people, or burglary can be detected with monitoring an alarm system associated with the MTC devices in place. The electrical devices that are not in use can be turned off aromatically to save the energy. Moreover, to reduce the power consumption costs, electronic devices can be used at the time of the day that the energy price is lower.

Smart city is another concept, which has attracted a lot of attention recently. For example, in a smart city, advertisements can be delivered to a customer according to his/her preference or hobby. A customer would be notified about the store in the nearby area that is selling his/her wanted product.

Furthermore, the MTC technology can help to optimize inventory, handle payment services and provide automatic updates on maintenance needs, leading to reduced costs while meeting the customers' needs.

1.2.4.2   Smart Lighting

Smart lighting systems can be used for homes, offices, and streets to improve the energy saving. As the urban population grows rapidly, highly efficient lighting systems can lead to reduction in carbon emission. For example, MTC technologies enable remote street light control allowing the M2M user applications such as the city lighting control managers to monitor and control street lights by smart phones, turning them on or off automatically depending on local illumination levels and traffic intensity.

1.2.5   Security and Public Safety

MTC technology is a perfect choice to provide security for private residential, commercial and public locations. In particular, it can provide cost-effective, rapid, and flexible deployment for remote surveillance, remote burglar alarms, personal tracking, and public infrastructure protection. These applications are described below.

1.2.5.1   Remote Surveillance

Remote surveillance is used to monitor open areas, valuable assets, people or pets for appropriate protection, where M2M sensors are deployed in video cameras to transmit signals either continuously or periodically. More specifically, the M2M application can help to detect possible risky situations, trigger proper actions, alert authorities, and keep an eye open to all suspicious activities and incidents. Examples are to inform the user if a specific object has been moved to/from a restricted area, report to an unauthorized entity, and provide the exact locations of the incidental events. In this scenario, the event has to be notified immediately to the owner, authorities, and/or to the security companies.

1.2.5.2   Personal Tracking

In these applications, persons are equipped with MTC devices, and optionally a GPS function to transmit the information regarding the location of persons automatically or on demand to a server application which monitors the positioning status of the intended person. Typical use of these applications are for health care, elderly or child monitoring.

1.2.5.3   Public Infrastructure Protection

Another application of MTC technology is in public infrastructure protection which is about monitoring and protection of various infrastructures including roads, bridges, tunnels, buildings, cables, and pipes. These applications help the government to reduce the infrastructure maintenance cost and enhance the operational efficiencies. By equipping the infrastructures with sensors and RFID tags, the monitoring and maintenance can be done easily.

1.2.6   e-Health

In this application area, various scenarios exist such as tracking or monitoring a patient or a segment of an organ in a patient, identification and authentication of patients, diagnosing patient conditions and providing real-time information on patients health related data to the remote monitoring center. One of the applications of MTC technology in health care systems is identification and authentication which are used in a variety of forms. For example, they are used to reduce the risk of wrong treatments to patients, provide real-time-based electronic medical record and also privacy protection against possible medical data leakage. In addition, they are used to grant security access to restricted areas and containers.

1.3   Machine-type Communication Characteristics

Machine-type communications have different traffic characteristics than the traditional networks dedicated to human-to-human (H2H) communications, in which a higher demand is on the downlink. In contrast to H2H communications, in MTC, since there are many applications for data gathering and reporting purposes, the uplink traffic originated from devices to the AP is heavier. Furthermore, as devices may transmit packets sporadically, the assumption of saturated scenarios, i.e., devices always have data for transmission, is no longer valid in these networks. For instance, in a smart metering application, the device only transmits a packet if a power outage happens or in a thermal monitoring application devices measure the temperature periodically but only transmit a new packet if a variation has occurred between the last two measurements.

Furthermore, in these networks, heterogeneity is inevitable as devices might belong to different applications and report different events or measurements. This characteristic necessitates a wireless network infrastructure with ability to support multiple concurrent applications. Otherwise using the domain-specific approach for networks results in redundant deployments and underutilization of resources. To realize such an infrastructure, a virtualization framework needs to be used, allowing multiple services and applications to access deployed network infrastructure and share radio resources. Such framework helps to reduce network deployment expenses and improve resource utilization by partitioning the existent physical network resources in an efficient manner [6].

These characteristics of machine-type communications have impacts on all the layers of the network including the medium access control (MAC) layer, which is primarily responsible to provide multiple access schemes for sharing the medium among devices in a network. In the following, we first categorize multiple access schemes and then we discuss the characteristics that they need to possess for machine-type communications.

1.4   Multiple Access Schemes and their Requirements in MTC

Multiple access schemes can be broadly classified as scheduling-based and random access schemes. In scheduling-based schemes which can be further divided into request-based and non-request-based schemes, radio resources are allocated to the devices by a central entity. More specifically, in request-based schemes, devices need to transmit a request to the central entity, while in non-request-based schemes, resources are assigned to devices without receiving any request from devices. Different from scheduling-based schemes, in random access schemes devices compete with each other to access the channel.

In MTC systems, one of the main requirements expected from the access schemes is high *throughput*. In fact, as wireless resources are scarce, it is desirable to maximize their efficiency. Using non-request-based scheduling schemes can lead to underutilization for an unsaturated network, where devices may not have any packet to transmit in their assigned time-slots. On the other hand, request-based scheduling schemes suffer from signaling overhead. Specifically, in MTC, where the size of the data packets can be small, the signaling overhead can have considerable negative impact on the throughput performance. Furthermore, in random access schemes, collision is unavoidable which affects the throughput performance.

Another key consideration for access schemes in MTC is *scalability* or *massive connectivity*. As in these networks a large number of devices may coexist in a single cell, it is important that the access scheme can support massive connectivity. Furthermore, the network conditions may vary over time, thus, the access scheme should be able to adjust its configuration in such dynamic environment without requiring significant control information exchange. Random access schemes like CSMA do not require information exchange; however, with increasing number of devices, their performance degrades due to collisions. On the other hand, request-based scheduling schemes require significant resources for the handshake procedure. Therefore, in scenarios with a large number of devices, their performance might be low [7].

Furthermore, for some applications, the *latency* is a critical factor that should be catered. For instance, intelligent transportation systems with autonomous driving, industrial process automation systems, and e-health applications, have strict demands on both reliability and timing. It is clear that random access schemes cannot guarantee the requirement of these applications, as the device with an urgent packet to transmit has to compete with others to access the channel. In addition, the transmitted packet may fail due to collision. On the other hand, request-based scheduling schemes also require exchange of messages between the device and the AP before the transmission happens, which might cause a noticeable delay. Thus, the MTC access scheme should be able to support these types of applications as well.

Moreover, using a virtualized wireless network brings its own challenges. In these networks, a single physical infrastructure and radio resources are partitioned among different applications or so-called slices. In this partitioning, which is also known as resource slicing, it is important to ensure that different slices are well isolated so that any change in one slice, such as variation in the number of devices or fluctuation of channel status, does not affect the resource allocation for other slices [8]. Another aim of using a virtualized framework is to gain high utilization of resources. The problem is that these two requirements of wireless virtualization, i.e., efficient resource utilization and isolation, are conflicting specifically in uplink transmission as traffic is generated at devices, while resources are allocated by the AP. For instance, in order to provide strict isolation among slices, an exclusive time-share can be reserved for each slice based on its requirement. However, such allocation could lead to underutilization for an unsaturated network, where devices may not have any packet to transmit in their assigned time-slots.

Last but not least, as the licensed bands are getting crowded, a substantial fraction of the machine-type networks is expected to operate in the unlicensed bands. In these scenarios, problems such as interference and unfair sharing of the bandwidth between the coexisting networks over the unlicensed bands may arise which should be addressed in the MAC layer [7].

1.5   Book Organization

The primary objective of this work is to discuss effective access schemes to meet the requirements of machine-type communications. In particular, we concentrate on achieving high spectral efficiency, satisfying quality-of-service (QoS) requirements of network slices, supporting massive connectivity, and enabling coexistence of different wireless technologies in these environments. In the following, the organization of this book is provided.

In Chap. 2, we review relevant multiple access schemes in machine-type communications that are useful for the development of MAC schemes in the subsequent chapters.

Considering the dynamic environment of machine-type communications, in order to improve the channel utilization, the traffic statistics information could be leveraged to efficiently configure a MAC scheme adapting to varying conditions. Thus, in Chap. 3, we propose a traffic-aware CSMA with deterministic backoff. In this scheme, with the aid of deterministic backoff values, collision is avoided. Moreover, since backoff values are assigned by the AP to the devices, QoS per slice can be considered.

The above approach performs very well for small to medium-sized networks for almost saturated devices, however its performance decreases for networks with large number of devices. Moreover, it may lead to starvation for devices with low probability of packet transmission. Thus, adopting a fixed MAC scheme cannot meet optimal characteristics along multiple dimensions. Furthermore, in practice there might not be any prior knowledge of traffic statistics or the statistical parameters might change over time. Therefore, employing an appropriate learning algorithm is crucial to acquire the traffic statistics such that the expected total throughput is maximized. In other words, such learning algorithm targets at mitigating the regret defined as the difference between the throughput obtained by the optimal solution for unknown traffic statistics and the achievable throughput with a priori knowledge.

To address the aforementioned issues, in Chap. 4, we present a reconfigurable access scheme which switches from contention-free to contention-based access regime, adaptive to the updated traffic statistics. The logic behind this scheme is assigning devices with high probability of packet transmission to the contention-free regime and allowing the rest of devices to compete in the contention-based regime. In particular, we propose the optimal scheduler that determines the partition between the two regimes. We also propose a scalable solution for a large number of devices as in M2M networks with considerably less computational complexity. In particular, to overcome the computational burden caused by a large number of devices, the optimization problem is transformed using approximations for random access throughput and airtime. Subsequently, we propose an efficient iterative algorithm to solve the approximated optimization problem, where each iteration is decomposed into two sub-problems: one belongs to the linear-programming category, and the other is of the difference of convex (DC)-programming type.

In addition, in Chap. 5 we develop a reinforcement learning algorithm based on Thompson sampling (TS) for scenarios of unknown packet arrival probabilities. We analytically prove that the proposed Thompson sampling-based learning algorithm can efficiently balance the trade-off between exploration and exploitation and achieves the optimal regret bound.

In Chap. 6, we assume a scenario that devices belong to two different access networks, i.e., one network uses a contention-free approach while the other one employs the contention-based scheme. An example for this scenario is LTE-WiFi coexistence over unlicensed band, where schedule-based and random channel access are employed by the LTE and WiFi, respectively. To maximize the network throughput, we propose a coordinated structure for coexistence of these two networks over the unlicensed spectrum. In such a coordinated model, the control of spectrum access between the two systems are governed by a virtualized network entity. This entity manages the channel access between these two systems such that the overall spectrum efficiency is improved, while the WiFi performance does not fall below a certain level. The corresponding optimization problem is formulated and an iterative algorithm is developed to find the optimal solution using complementary geometric programming (CGP) and monomial approximations. Furthermore, aiming to address QoS assurance for LTE devices, we obtain an upper bound for average delay of these devices. This analysis could be a basis for admission control of LTE devices in unlicensed bands.

Moreover, aiming to further increase the network throughput, in Chap. 7, we propose a non-orthogonal multiple access scheme in which multiple devices can successfully transmit over the same time-slot. In particular, in the proposed scheme, devices with high probability of having non-empty queue and sufficient channel gain differences are paired and assigned a single time-slot, while the rest are considered for the random access regime. An efficient device-pairing scheme for uplink scenarios is proposed, where by applying successive interference cancelation schemes, the AP is able to decode the signals received from each of the devices.

In Chap. 8, to minimize the signaling overhead, a distributed learning-based access scheme is developed in which each device independently adapts its transmission length to the optimal values over time by learning the number of active devices based on locally available information.

Finally, Chap. 9 concludes this book.

References

1. A. Biral, M. Centenaro, A. Zanella, L. Vangelista, M. Zorzi, The challenges of M2M massive access in wireless cellular networks. Digital Commun. Netw. **1**(1), 1–19 (2015)
2. H. Shariatmadari, R. Ratasuk, S. Iraji, A. Laya, T. Taleb, R. Jäntti, A. Ghosh, Machine-type communications: current status and future perspectives toward 5G systems. IEEE Commun. Mag. **53**(9), 10–17 (2015)
3. S.K. Sharma, X. Wang, Towards massive machine type communications in ultra-dense cellular IoT networks: current issues and machine learning-assisted solutions. IEEE Commun. Surv. Tutorials **22**(1), 426–471 (2019)
4. F. Ghavimi, H.-H. Chen, M2M communications in 3GPP LTE/LTE-A networks: architectures, service requirements, challenges, and applications. IEEE Commun. Surv. Tutorials **17**(2), 525–549 (2014)
5. P. Spiess, S. Karnouskos, D. Guinard, D. Savio, O. Baecker, L.M.S. De Souza, V. Trifa, SOA-based integration of the internet of things in enterprise services, in *IEEE International Conference on Web Services* (IEEE, New York, 2009), pp. 968–975
6. I. Khan, F. Belqasmi, R. Glitho, N. Crespi, M. Morrow, P. Polakos, Wireless sensor network virtualization: early architecture and research perspectives. IEEE Netw. **29**(3), 104–112 (2015)
7. A. Rajandekar, B. Sikdar, A survey of MAC layer issues and protocols for machine-to-machine communications. IEEE Internet Things J. **2**(2), 175–186 (2015)
8. M. Richart, J. Baliosian, J. Serrat, J.-L. Gorricho, Resource slicing in virtual wireless networks: a survey. IEEE Trans. Netw. Serv. Manag. **13**(3), 462–476 (2016)

References

Chapter 2
Multiple Access Schemes for Machine-Type Communications: A Literature Review

2.1 Multiple Access Schemes for MTC

In this section, we review the existing access schemes in the literature for MTC, categorized into scheduling-based, random access, and hybrid schemes, which combine both scheduling-based and random access schemes.

2.1.1 Scheduling-Based Schemes in LTE

First, we introduce the LTE frame structure and then we present some existing scheduling-based access schemes proposed for MTC in LTE.

2.1.1.1 LTE Frame Structure

In LTE, downlink and uplink transmissions are organized into frames that are 10 milliseconds (ms) long. Each frame is split into ten 1 ms sub-frames and each sub-frame consists of two 0.5 ms slots, where each slot is composed of multiple symbols. In the frequency domain, the total bandwidth is split into units of subcarriers, where a unit of 12 subcarriers for a duration of one slot is denoted as a resource block (RB).

Different parts of LTE frames are assigned for different purposes known as physical channels and signals. In the uplink frames, there are three physical channels: physical uplink shared channel (PUSCH), physical random access channel (PRACH), and physical uplink control channel (PUCCH). PUSCH is used to carry uplink data while PRACH is used by devices to send resource access requests. Finally, PUCCH is used by devices to send uplink control information to the BS

© Springer Nature Switzerland AG 2020
T. Le-Ngoc, A. Dalili Shoaei, *Learning-Based Reconfigurable Multiple Access Schemes for Virtualized MTC Networks*, Wireless Networks,
https://doi.org/10.1007/978-3-030-60382-3_2

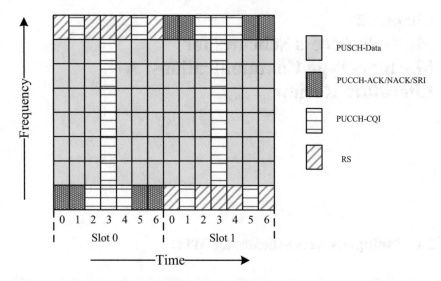

Fig. 2.1 Physical uplink channels in an LTE sub-frame [2]

such as scheduling request and buffer status report (BSR) [1, 2]. The position of these channels in an uplink sub-frame are shown in Fig. 2.1.

2.1.1.2 Existing Scheduling-Based Access Schemes for MTC Over LTE

In LTE, the uplink scheduling is done at the BS, and the allocation decisions are passed to the devices over appropriate control channels. In particular, in the uplink LTE, if a device wants to transmit some data, it needs to send a request to the BS. The process to send the request is called scheduling request (SR) procedure, in which a resource in the PUCCH channel is periodically allocated to the device by the BS. The device with some data to transmit, sends its request in the allocated resource and the requested resource will be allocated to the device [3].

In the literature, different priority metrics and approaches are used for scheduling MTC devices. One of the widely used scheduling metrics is data transmission deadline. For example, in [4], a scheduler is designed aiming at maximizing the percentage of uplink packets that their delay requirements are met. To realize this goal, a new element to the data packets is added which allows devices to inform the BS of the age of their oldest packet in their queues. The BS uses this information to calculate the emergency metric for each scheduling request which depends on the time remaining to the deadline of the packet as well as the amount of pending data in the device queue. To perform the scheduling, the BS ranks the requests according to the emergency metric and assigns radio resources to devices having the highest values of the emergency score.

The study in [1] also employs the metric-based approach for scheduling MTC devices. Targeting scenarios devices report their collected data to the BS in a periodic manner, higher priorities are assigned to devices where their data packets contain dissimilar information with respect to the previously sent data. More specifically, the aim of this scheme is to use radio resources for transmitting important information rather than redundant data. Each device with a packet for transmission computes its statistical priority score and sends the score in the scheduling request. In the scheduling process, the scores are taken into account such that requests with higher scores are given higher priorities.

In an another approach, to schedule MTC devices, grouping schemes are used where devices are clustered into QoS classes and radio resources are allocated accordingly. This approach is applied in [5], where the BS keeps a separate queue for each QoS group to store their corresponding scheduling requests. For each group, the number of packets served in a unit time, is dynamically adjusted such that the overflow probability of the queue, i.e., the probability that the queue length becomes larger than a certain value is kept lower than a given threshold.

The grouping approach is also used in hybrid M2M/H2H networks where scheduling techniques should efficiently accommodate both traditional H2H traffic and the MTC device traffic with different QoS requirements [6, 7]. In [6], first, resources are split between H2H users and MTC devices, then the share of MTC devices is further divided such that a balance between throughput maximization and meeting the delay requirements of MTC devices is provided. In other words, the authors use both throughput and delay as a metric to schedule MTC devices. On the other hand, in [7] the aim of the MTC scheduler is to ensure fairness in allocation of resources to devices. To that end, the BS assigns higher priorities to the devices that were allocated fewer resources over time and that have shortest remaining time to exceed the maximum tolerable delay.

In general, the proposed scheduling-based schemes work well for a low number of users/devices. However, as LTE assigns each user/device one PUCCH, in the presence of a large number of devices, a shortage of PUCCH resources is possible. Moreover, these schemes are not proper for scenarios devices generate small-size packets in a sporadic manner, as the amount of resources used for the request-grant procedure compared to the transmitted data is large.

2.1.2   Random Access Schemes in LTE

In addition to scheduling-based schemes, LTE also provides a random access (RA) procedure for uplink transmissions. In the following, first, the RA procedure is described. Then, the reasons that RA procedure may fail are provided along with its drawbacks in MTC systems. After that, some proposed enhancements to the RA procedure are discussed.

2.1.2.1 Random Access Procedure in LTE

The RA procedure (Fig. 2.2) consists of four steps: (1) RA preamble transmission from the device to the BS (Message 1), (2) RA response (RAR) from the BS to the device (Message 2), (3) connection request message from the device to the BS (Message 3), and (4) connection resolution message from the BS to the device (Message 4).

In the first step, a device wanting to initiate an RA attempt, randomly chooses one RA preamble. RA preambles are orthogonal bit sequences generated by cyclically shifting a root sequence. After generating the preamble, the device sends it to the BS in the first available RA slot. In the frequency division duplex (FDD) operation, an RA slot consists of 6 units in the frequency domain, while it can occupy 1, 2, or 3

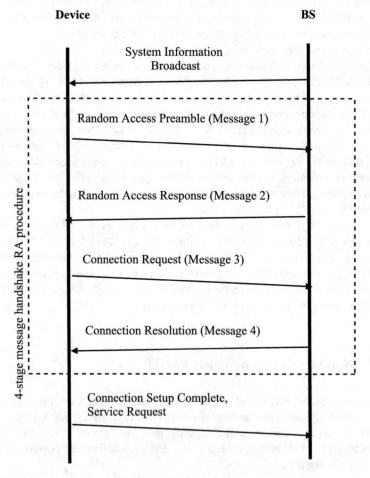

Fig. 2.2 Contention-based random access in LTE

sub-frame(s) in the time domain depending on the preamble format. At this step, the device just transmits the selected preambles and not the device ID. If the preamble is only chosen by one device, the BS is able to decode it. However, if multiple devices choose the same preamble, the BS may be able to detect the collision, if the devices are at the different distances from the BS. In the case of successful decoding of the preamble, in the second step of the RA procedure, the BS transmits the random access response (RAR) which includes the uplink resource allocation for transmitting the third message of the procedure. In the third step, the actual RA message (i.e., radio resource request, scheduling request) is transmitted to the BS. This implies that if multiple devices chose the same preamble and collision was not detected by the BS, they transmit the scheduling request over the same resource which causes collision. Finally, in the last step, if the BS receives the scheduling request, it sends a contention resolution message as a reply to the third message [8–10].

2.1.3   Failure of RA Procedure and Its Inefficiency in MTC Systems

In the aforementioned RA procedure, after sending the preamble in the RA request (Message 1), the device sets an RAR window and waits for the BS's response with an uplink grant (Message 2) in the RAR message. If the device successfully receives its Message 2 within the defined RAR window, the device sends the radio resource control (RRC) connection request (Message 3) to the BS. At this stage, the device starts the Message 4 timer and waits to receive its own RRC connection setup message (Message 4) from the BS [11].

The RA procedure may fail due to the following reasons.

- Preamble transmission failure: As mentioned above, a transmitted preamble may be ignored by the BS due to the RACH preamble collision (two or more devices transmit the same preamble at the same time). In addition, preamble transmission failure may occur if the preamble transmission power is insufficient.
- Message 2 reception failure: If there is no sufficient downlink radio resources (PDCCH), the BS fails to transmit the RAR (Message 2).
- Message 3 transmission failure: To transmit Message 3 to the BS, a device applies hybrid automatic repeat request (HARQ). If the device fails to send Message 3 to the BS by this method, it fails in Message 3 transmission.
- Message 4 reception failure: The HARQ method is also used by the BS to send Message 4 to the device. Failure in Message 4 reception happens, if the device does not receive Message 4 before Message 4 timer expiration either due to insufficient PDCCH resources or an imperfect channel condition.

A RACH trial indicates the action to perform the four-step RACH procedure once, and a RACH trial is considered a failure if any of the aforementioned failures occurs. If a RACH trial failure occurs to a device, the next RACH trial is done by the device after waiting for a random backoff period. The backoff counter is uniformly selected from $[0, W]$, where W is denoted by the backoff indicator (BI). A device performs at most K RACH trials, and if it encounters K RACH trial failures, the RACH procedure is failed.

The total number of preambles is 64 and they are divided into two groups: contention-free RA preambles and contention-based RA preambles. Some preambles are reserved for contention-free access which is needed for high-priority services for handover, while the rest are used for contention-based RA. Assuming the number of preambles reserved for contention-free RA is 10, and total number of access opportunities per second is 200, then there is a capacity of 10800 preambles per seconds in the absence of collisions. However, as collisions are inevitable, the maximum capacity limit cannot be reached. The performance of the RA is even worse in massive MTC systems due to high probability of collision caused by concurrent massive access requests. One approach to lower the physical RACH load is to assign more resources for RA access in a frame, however, this approach reduces the amount of resources required for data transmission. Therefore, the tradeoff between the allocated resources for data transmission and access opportunities per frame should be balanced.

In general, due to the following reasons, the RA procedure employed in LTE systems is not efficient for MTC [11].

1. The number of preambles is limited, therefore in massive MTC scenarios, the massive number of simultaneous transmissions of the same preambles causes high collision probability and consequently high access failure rate and access delay.
2. For large number of access requests, additional downlink resources are needed as each RAR message for one MTC device has 56 bits.
3. The signaling overhead in RA procedure is large for MTC, as the size of packets in these systems is small. Therefore, even if a device successfully transmits its data using the RA procedure, the efficiency of the scheme would be low.

2.1.3.1 Enhancements of the RA Procedure

To enhance the performance of the RA procedure, several solutions are suggested. For example, in [12–14], to reduce the congestion in a large scale network, the access class barring (ACB) method is proposed. In this scheme, the ACB factor p_{acb} is broadcast by the BS to the devices, where each device sends the preamble with probability p_{acb}, otherwise it defers its access time. In [12], a Markov chain-based traffic-load estimation scheme is proposed to adjust p_{acb} according to the estimated traffic load. In [13], it is also assumed that the number of devices applying for the preamble at each time-slot is unknown by the BS, and it proposes two Bayesian algorithms to estimate the number of active devices based on the number of idle preambles.

More specifically, for each time-slot t, first, the estimated average number of active devices, denoted by $\hat{N}_a^t$ is calculated. Then, based on this value, the optimal p_{acb}^t is calculated and broadcast to devices. It is shown that for N_m preambles, and N_a^t active devices, the optimal p_{acb}^t is

$$p_{acb}^t = \min(1, N_m/N_a^t) \tag{2.1}$$

Upon receiving p_{acb}^t, devices react and the BS observes the number of idle preambles denoted by N_i^t. Then, based on N_i^t, the average number of active devices is re-estimated using the a posteriori probability denoted by $\mathbb{P}(N_a^t|N_i^t, \hat{N}_a^t)$. If the new estimation denoted by $\hat{N}_a^{t'}$ is greater than $\hat{N}_a^t$, then it is interpreted as an increase in the number of active devices due to new arrivals. Subsequently, the newly activated devices are estimated as $N_n^t = \max(0, \hat{N}_a^{t'} - \hat{N}_a^t)$. Furthermore, the number of successfully transmitted connection request messages, denoted by N_c^t is considered as deactivated devices. Finally, N_a^{t+1} is estimated as

$$\hat{N}_a^{t+1} = \hat{N}_a^t + N_n^t - N_c^t. \tag{2.2}$$

To bring further performance improvements to the ACB method, a refinement of this scheme, called extended access barring (EAB) is proposed [15]. In this method, when the network is congested, only some of the devices are allowed access while the rest are barred from accessing the network. The list of the barred devices along with the duration of barring time is announced by the BS. The work in [16] evaluates the performance of this method. In order to provide better access prioritization for different traffic classes, several approaches are proposed [17–19]. The work in [17] proposes to use different ACB factors for different classes. Furthermore, in [18], to guarantee QoS, different number of PRACH slots are assigned to different classes. Different from [18], in [19] to support access prioritization, the set of preambles are divided between classes, where the number of preambles allocated to each class is dependent on the traffic load and the priority tuning parameter of the class. The drawback of the work in [19] is that the proposed method is for persistently high traffic loads, while in M2M networks devices with both persistent and non-persistent traffic loads may coexist. Furthermore, although it increases the access success probability under a relatively heavy load, but the access delay of MTC devices is severely degraded.

Another approach to address the RA congestion problem is to use dynamic allocation of RACH. In this approach, the BS allocates resources in frequency domain, time domain, or both based on the RA congestion level. It is obvious that the more resources are assigned to the RA access opportunity, the less would remain for data transmission. The performance of this approach is investigated in [20], which recommends it as the primary solution to the RA congestion problem for massive MTC.

In [21], to address the limited number of preambles, an RA model based on the capacity approaching analog fountain code (AFC) is proposed. In this model, preambles are divided into multiple groups based on the delay requirements of devices. Each device initiates the RA procedure by selecting a specific RA preamble according to its delay requirement from the corresponding group of RA preambles. It is assumed that uplink power control is performed by devices such that the average received SNR from all devices at the BS are the same. Therefore, in the second step of the RA procedure, the BS is able to detect the number of devices that have selected the same RA preamble. This information is broadcast in RAR to all contending devices for each of the detected preambles. In the third step, each device that sent the same preamble, receives the information about the total number of devices using that specific preamble. Based on this information, the device first calculates the length of a random seed and then generates an orthogonal random seed and transmits it to the BS. After the RA procedure, transmissions start. More specifically, devices that have selected the same preamble, transmit their AFC coded symbols in the same RB with the same access probability, but with different random seeds. As both the BS and the device use the same random seed, the BS is able to decode the received coded symbols. The proposed RA procedure is shown in Fig. 2.3.

In [22], a collision-resolution-based RA model is proposed to resolve the collisions occurred in the RA procedure. In this model, RA preambles are split between HTC and MTC, and the collision resolution technique is only used for MTC. In the RA procedure, a device randomly selects a preamble, if the selected preamble is chosen by other devices as well, collision occurs. To resolve the collision, the BS allocates a set of new preambles to the collided MTC devices and transmits this set in the RAR, while if devices belong to HTC, the BS does not transmit the RAR and the collided devices perform the next RA trial. The MTC devices use the new preamble set to retransmit the RA request. If they collide again, another set of preambles is assigned to them. This process continues until the preamble of each device is detected by the BS. An illustration of this RA procedure is provided in Fig. 2.4. In this splitting binary tree, the root is denoted as level 0. For each collision at level 0, a set of m preambles is reserved for level 1. Similarly, m preambles are reserved at level 2, for each collision occurred at level 1 and the process continues until all collisions are resolved. It should be noted that in this model, the number of preambles in each reserved set (m) is dynamically adjusted according to the collision rate.

In general, in LTE, random access schemes can outperform scheduling-based schemes for machine-type traffic, however, the proposed RACH procedure still has four message-exchange steps, leading to noticeable signaling overhead.

2.1.4   Random Access Schemes in WiFi

In WiFi, to access the channel, CSMA is used by devices. In this scheme, prior to the transmission, the device has to sense the channel and transmission only happens if the channel is detected as idle. This scheme works well for a small number of

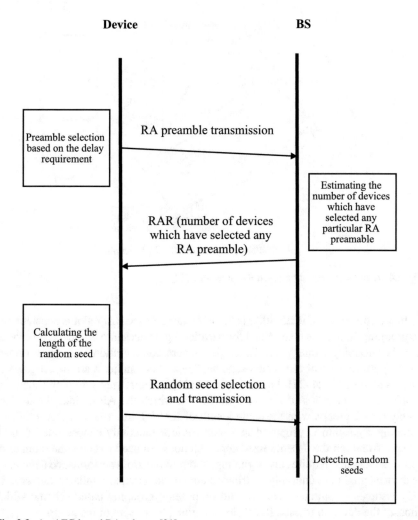

Device **BS**

Preamble selection based on the delay requirement

RA preamble transmission

Estimating the number of devices which have selected any particular RA preamable

RAR (number of devices which have selected any RA preamble)

Calculating the length of the random seed

Random seed selection and transmission

Detecting random seeds

Fig. 2.3 An AFC-based RA scheme [21]

active devices, however, as the device density increases, its performance in terms of throughput and delay degrades quickly due to collisions, which makes the scheme unsuitable for MTC.

To support communications among MTC devices, a new amendment of this scheme is proposed by IEEE 802.11ah [23]. In this scheme, devices contend with each other to access the channel, however, the number of devices that contend at the same time is restricted. More specifically, in this standard, the channel time is divided into beacon intervals, each of which is further partitioned into multiple restricted access window (RAW) slots. Devices are grouped and a RAW slot is assigned to each group. That is at each RAW slot, only a group of devices can contend to access the channel.

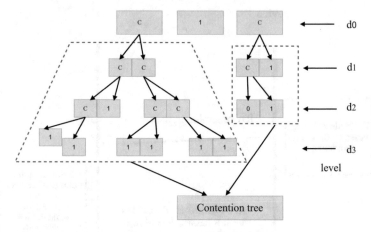

C=Collision, 1=Success, 0= No detection

Fig. 2.4 A collision-resolution-based RA procedure [22]

In the conventional IEEE 802.11ah, the duration of the RAW slot is constant and no grouping strategy is determined for partitioning devices into groups. It is obvious that the grouping strategy and RAW parameters configuration have huge impact on the performance of the access scheme. Thus, these subjects are investigated in various studies [5, 24–27]. For example, in [24], a centralized grouping scheme is proposed for scenarios of saturated devices in which the AP divides the devices evenly into k groups with the same length of RAW. Furthermore, a decentralized grouping algorithm is proposed in which devices randomly choose one of the k groups. Although the decentralized approach does not need to know the number of active devices and requires less signaling, it may suffer from performance deficiency as devices might get unevenly distributed among the groups. Similarly, the work in [25] assumes a saturated scenario and proposes a grouping-based scheme which chooses the duration of each RAW slot according to the size of the group.

The work in [28] considers a scenario in which devices generate periodic packets with different rates and lengths. Assuming that the beacon interval is evenly divided into a fixed number of RAW slots, the problem of assigning devices to the RAW groups is formulated as an optimization problem in which the objective is to maximize the network throughput. It is shown that the proposed optimization problem is NP hard, therefore, to address the computational complexity, a greedy algorithm is developed to solve the problem. The proposed greedy algorithm is iterative, where at each iteration, first, the group having the lowest utilization is selected and then the device which provides the largest increment to the utilization of the selected group is added to that.

In [26], it is assumed that the beacon interval consists of one RAW for the downlink access and one RAW for the uplink traffic. To determine the length of the uplink RAW, an algorithm is proposed, which estimates the number of uplink devices. The main drawback is that considering only one RAW for the uplink access may lead to a huge number of collisions and consequently low throughput. Furthermore, the derivation is based on the number of devices, while in M2M networks, devices may be unsaturated.

The work in [27] focuses on the design of grouping-based MAC protocols for an event-driven scenario, in which a smart meter is attached to each electric vehicle and is used to report the charging parameters to the network. However, the results are derived for a single application, in which all devices have the same traffic parameters. Considering a network running a single application might be unrealistic for the future network as different applications may share the same infrastructure.

Although using a proper grouping-based strategy improves the WiFi performance, the throughput cannot be maximized due to collisions. In the following, we review some approaches which are proposed to either eliminate the occurrence of collisions or use random access schemes only for sending the scheduling requests.

To eliminate collisions, [29] proposes the deterministic backoff (DEB) method. In this scheme, each device is assigned a unique backoff value by the AP, where backoff values are transmitted in a single beacon. For instance, if there are N_d devices in the network, the AP assigns backoff values from 1 to N_d to different devices in a round-robin fashion. This information is broadcast by the AP to the devices in the beacon.

To access the channel, the device senses the channel at each time unit starting from the first time unit of the frame. If the channel is sensed idle, the backoff value counter is decreased by 1. Otherwise, the device freezes the backoff value counter and continues decrementing when the channel becomes idle again. Finally, the device transmits its packet when the backoff value counter reaches zero. In fact, this approach acts as a virtual polling. In a polling-based MAC, the AP transmits a polling packet to the intended device whenever it wants to receive/transmit a packet from/to that device [30]. The polling-based mechanism has its own disadvantages and suffers from additional signaling overhead at each polling packet. Moreover, if the device cannot receive the polling packet, it would miss its transmission opportunity. However, using DEB, the need for the polling packet exchange is eliminated via carrier sensing.

In [29], a saturated traffic condition is considered and thus deterministic backoff values are assigned to different devices in a round-robin manner. The benefit of using this scheme compared to the scheduling-based schemes is that it allows devices to transmit packets with variable lengths. However, such an approach is not suitable for unsaturated traffic scenarios, since the time wasted for backoff purposes reduces the resource utilization.

2.1.5 Hybrid Schemes

There are some works that propose to combine RA and scheduling-based schemes to address the requirements of MTC. In [31], a CSMA-time division multiple access

(TDMA) hybrid scheme is proposed, in which time is divided into superframes. In particular, each superframe consists of four parts: notification period (NP), contention only period (COP), announcement period (ANP), and transmission only period (TOP). The superframe starts with the NP, which is used by the BS to announce the beginning of the COP to all devices. During the COP, devices with data use p-persistent CSMA to send transmission requests to the BS. Successful devices are allocated time-slots to transmit data in the TOP and devices are informed of their time-slots during the ANP. The length of the COP may vary from superframe to superframe. An optimization problem is solved by the BS to determine the optimum COP length and the number of devices that are allowed to transmit in the TOP. The length of the COP as well as the optimum contention probability for the p-persistent CSMA are broadcast to all devices by the BS during the NP. This scheme is extended in [32] in which QoS provisioning and fairness are also considered. In order to reach these goals, [32] allows devices to choose their contention probabilities according to their priority and observed throughput.

In [33], the authors propose an access scheme for MTC targeting the scenarios of both periodic and non-periodic traffic. In that work, devices are divided into groups such that the probability that a collision happens during the corresponding allocated time is below than a certain threshold. Devices belonging to the same group contend with each other to transmit the access request. The drawback of this approach is that, similar to [32], a separate phase is dedicated for request gathering. Furthermore, it is assumed that the traffic parameters of the devices are known.

An alternative solution to address the requirements of MTC is to deploy the distributed queuing (DQ) mechanism. In this scheme, the frame structure is divided into three parts: (1) C_{DQ} sub-slots for collision resolution, (2) one slot for data transmission and (3) one sub-slot for transmission of feedback information from the AP to devices [34]. The access scheme works based on two queues: contention resolution queue (CRQ) and data transmission queue (DTQ). At each frame, the devices at the front of CRQ, randomly choose one of the C_{DQ} contention sub-slots to transmit an access request sequence (ARS). Thus, the status of each sub-slot can be (1) idle (no ARS is transmitted), (2) successful (only one ARS is transmitted), and (3) busy (more than one ARS is transmitted). The AP broadcasts that information at the end of the frame in a feedback slot. Devices with successful ARS transmission are added to the DTQ, while colliding devices over each sub-slot are added to CRQ. Furthermore, at each frame, during the data transmission phase, the device at the front of DTQ transmits its packet. It should be noted that each device can compute its position in each queue based on the feedback information sent by the AP. It has been shown that this scheme can achieve better performance compared to the standard ACB method [35]. However, for small packet sizes, the overhead of the collision resolution part might be large.

2.2 Machine-type Communications in LTE/WiFi Coexistence

So far, we have only discussed access schemes proposed for MTC systems in which all devices access the channel through the same wireless technology. However, to improve the spectral efficiency, the same channel may be shared between different wireless technologies. As mentioned in Chap. 1, LTE operation on unlicensed bands, so-called unlicensed LTE (U-LTE), is considered by the third-generation partnership project (3GPP) as a promising solution to meet the growing wireless data demand and to improve the spectrum efficiency. Although transmission across both unlicensed and licensed bands can boost LTE, such an approach may jeopardize the performance of WiFi systems solely operating on unlicensed bands for data transmission. The reason is that LTE networks exploit a scheduling-based channel access, while in WiFi a contention-based scheme is applied, in which the device would randomly access the channel once it is detected idle. Therefore, in a coexistence scenario that both systems share the same channel, starvation may happen for WiFi as the whole airtime may be occupied by the LTE network [36–41].

In order to address this issue, two approaches are so far proposed, LTE-Unlicensed (LTE-U) and licensed-assisted access (LAA). In LTE-U, developed in 3GPP Releases 10/11/12, a duty-cycle-based approach is used in which at each duty cycle, LTE transmits over only a portion of a duty cycle, securing the rest of cycle for WiFi [42]. The main problem of this approach stems from no carrier sensing before LTE transmissions, as WiFi transmissions occurring in the LTE cycle might be interrupted by LTE transmissions. On the other hand, in LAA which has been featured in 3GPP Release 13, the LTE BS is equipped with the listen-before-talk (LBT) mechanism, i.e., carrier sensing is performed before any transmission [43–45]. In fact, in LAA, the LTE BS deploys a procedure which is similar to the random access scheme used by WiFi devices, therefore it is easier to achieve fairness between the two networks.

It should be noted that although WiFi DCF and LTE-LAA LBT share very similar random access structure, there exist a number of key differences, which are summarized as below.

- Access priority: In DCF, the backoff parameters are the same for all devices, thus it provides an equal opportunity for channel access. However, in LTE-LAA four different access priority classes are defined for different traffic/service types, where each class has its own transmission/backoff parameters. IEEE 802.11e also contains the class-based prioritized access categories.
- Transmission duration: In DCF, a WiFi device is allowed to transmit only a single packet when it wins the channel access. However, in LTE-LAA, the device can transmit for a TXOP duration of up to 10 ms normally, and 8 ms in coexistence mode.

- Channel sensing duration: In DCF, in order to access the channel, the device has to sense the channel idle for a distributed interframe space (DIFS) duration of $34\,\mu s$ which contains a short interframe space (SIFS) of $16\,\mu s$ and two time-slots with slot duration of $9\,\mu s$. In LTE-LAA, a device is required to perform a clear channel access (CCA) procedure using energy detect mechanism over a defer period of $16\,\mu s$ followed by multiple time-slots with slot duration of $9\,\mu s$, depending on the access priority of the device.
- Backoff parameters: WiFi and priority classes in LTE-LAA have different initial backoff window sizes and the maximum number of backoff stages.

2.2.1 Existing Works on LTE-LAA and WiFi Coexistence

In the literature, the LAA approach for LTE and WiFi coexistence (Fig. 2.5) has been investigated under different conditions.

Assuming *saturated* LTE and WiFi networks, in [46], an analytical work is presented to provide proportional fairness between two networks. To reach this goal, it is assumed that LTE transmits with probability of q_l in bursts of duration T_l and these parameters are optimized such that the proportional fairness can be achieved. More specifically, to obtain q_l and T_l, an optimization problem is solved in which the objective is to maximize the total throughput of WiFi and LTE while proportional fairness among LTE-U and WiFi devices is ensured. The work in [47] also uses the LBT-based method in which the aim is to guarantee the required rate of LTE devices while the collision probability of WiFi devices is minimized. In [45], assuming that the number of WiFi devices is unknown, a further analysis is

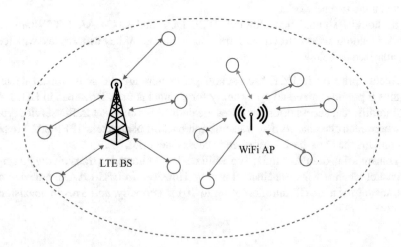

Fig. 2.5 Illustration of WiFi/LTE coexistence in unlicensed bands

presented to estimate the population of the WiFi system. Moreover, in [48, 49], the fair coexistence problem is addressed for multi-cell scenario comprised of multiple WiFi APs and LTE BSs.

The analytical performance of LTE-LAA and WiFi coexistence is studied in [50, 51]. However, in these works, to avoid overly complicated models, some simplifications are considered. For example, in [50], it is assumed that both WiFi and LTE-LAA have the same sensing period, while as mentioned above, the sensing time in these networks is different. In [51], it is considered that WiFi has one access priority class and LTE has four priority classes. However, in any specific scenario, all the LTE-LAA devices are assumed to belong to the same access priority to simplify the analysis.

In [52], also the proportional fairness in WiFi and LTE coexistence is addressed, in which the proportional fairness is achieved if all devices device achieve an identical fraction of time over the channel. To reach this goal, the main features of LTE-LAA and WiFi coexistence, i.e., initial backoff window size, number of sensing slots, maximum bakoff stages, retry limits and transmission opportunities for LTE and WiFi devices can be tuned. In the proposed scheme in [52], optimal initial backoff window size and the number of sensing slots of LTE-LAA are tuned to provide proportional fairness between LTE-LAA and WiFi. In particular, an analytical model is established for estimating LTE-LAA and WiFi throughput in coexistence in saturated conditions by using a Markov model.

Under assumption of *unsaturated* LTE and WiFi networks, in [53], an optimization problem is studied with the aim of maximizing overall throughput, while maintaining WiFi throughput. In [54], maximum allowable packet arrival rates of both LTE and WiFi are derived under which WiFi delay requirement is guaranteed.

Although the LAA approaches proposed in [45–47, 53, 54] may lead to enhanced performance for WiFi compared to LTE-U, the utilization still cannot reach the optimal point due to the lack of coordination between the two networks. Thus, an efficient network structure along with a proper MAC need to be designed. In [55], a hyper AP is introduced which is able to operate as both LTE BS and WiFi AP. However, that work considers saturated scenarios while MTC systems are mainly involved with unsaturated devices. Furthermore, MTC systems are uplink-centric.

For *uplink scenarios*, in [9], the authors propose an approach in which if an LTE device requests one resource batch consisting of a number of resource blocks, the BS allocates multiple resource batches to the device. Because it is assumed that each resource batch might be occupied by WiFi devices with a known probability. Therefore, assigning only one resource batch to the requesting LTE device, may lead to low expected throughput for the device. On the other hand, assigning multiple resource batches to one device may lead to low utilization for resource batches, as for example if all allocated resource batches are free, only one of them will be used by the requesting LTE device. To address this issue, it is proposed to share M_{rb} resource batches among a number of LTE devices. More specifically, in [9], to maximize the utilization of resource batches by LTE devices, having the probability that a resource batch is occupied by WiFi devices, the optimal number of devices sharing M_{rb} resource batches is derived.

In [56], both uplink and downlink traffic are considered, in which the goal is to improve the performance of LTE and WiFi coexistence in unlicensed band and provide the proportional fairness in terms of throughput between uplink and downlink. To reach this goal, an optimization problem is formulated which takes into account throughput and bandwidth constraints and for each direction, i.e., downlink and uplink, assigns devices to either the LTE BS or a WiFi AP. Furthermore, as number of devices varies over time, a machine learning-based algorithm is proposed to forecast the total number of devices in the next time-steps. In particular, the support vector machine (SVM) algorithm is used due to its low computational complexity and high prediction accuracy.

In addition, uplink resource allocation for U-LTE is addressed in [9, 57]. However, in these works, to access the channel, LTE devices should perform channel assessment, which requires modifications on the current deployed LTE scheme. Moreover, none of these works analyzes delay performance for LTE devices. Such analysis is indispensable to offer satisfactory experience for LTE devices. Thus, in order to address MTC requirements in LTE/WiFi coexistence scenarios, a proper architecture scheme along with a fair allocation algorithm are required.

2.3 NOMA-Enhanced Multiple Access Schemes for MTC

In the above sections, we have presented access schemes in which to have a successful transmission, a single radio resource should be only used by one device at a time. However, as mentioned in Chap. 1, there is an another category of access schemes called NOMA schemes, which allow multiple devices to transmit over the same radio resource simultaneously while the receiver is able to decode all received signals using successive interference cancellation (SIC) techniques. Figure 2.6 shows the comparison between orthogonal access schemes (OMA) and NOMA schemes.

2.3.1 Overview of Uplink NOMA

In uplink NOMA, multiple devices non-orthogonally transmit to the AP on the same radio resource. At the AP, these signals are received in a superimposed form, where they cause interference to each other. In order to decode these signals, SIC technique can be used, where to ensure successful SIC, the received signal power of the devices should be distinctive [58–60]. As the signal of each device experiences a distinct channel gain, the received power of devices are different at the AP, which makes it possible to use NOMA, even if all devices transmit at the same power level.

In the SIC technique, the AP first decodes the signal of the device with the highest channel gain, since it is likely the strongest at the AP. To do that, the AP treats the

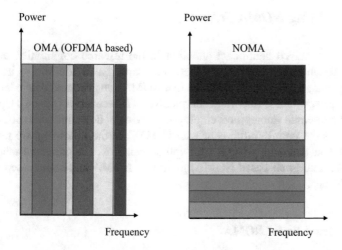

Fig. 2.6 Comparison of NOMA and OMA

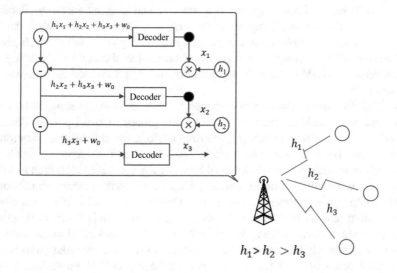

Fig. 2.7 Illustration of a 3-device uplink NOMA with SIC at the BS

signals from other devices as additive noise. After that, the decoded signal of the first device is subtracted from the received signal and the AP continues the decoding for the second device and treats the remainder as additive noise. This process is continued until all the devices are successfully decoded [61]. As a result, the highest channel gain device experiences interference from all devices and the lowest channel gain device effectively enjoys interference-free transmission if SIC is done without any errors. The procedure is shown in Fig. 2.7.

2.3.2 Existing NOMA Schemes

NOMA has received significant attention in the research community, due to its potential capability to improve the networks capacity. In addition to its spectral efficiency gain, it has been also shown that NOMA can accommodate a large number of devices, which is important for massive MTC scenarios [62–64]. Up to now, most of the research studies have employed NOMA for downlink scenarios [65–69]. However, as the uplink traffic is heavier in MTC, in the following, we review the existing works that deploy NOMA for uplink scenarios. The presented schemes are categorized into grant-based NOMA, grant-free NOMA and compressed sensing-based schemes.

2.3.2.1 Grant-Based NOMA

NOMA allows multiple devices to transmit over the same radio resource, but since it is an interference-limited system, it is not practical to simultaneously allocate a single radio resource to all devices of the network. Thus, in this scheme, devices are divided into multiple groups, where NOMA is exploited within each group, and among different groups, resources are allocated based on OMA. Evidently, the performance of NOMA is highly dependent on the algorithm used for grouping the devices [69].

In [58], to improve the network throughput, first, devices are grouped by using a sub-optimal algorithm and then for the given groups, optimal power allocation is derived. In the proposed scheme, to group devices, the channel gain differences among them are exploited. The work in [70] also uses the decomposition-based approach in which devices are clustered by using a graph-based algorithm and then power and bandwidth are optimized. In [60], a power control algorithm for uplink NOMA is proposed, where the outage probability and the achievable sum rate of the proposed scheme are theoretically analyzed. The work in [71] proposes optimal and sub-optimal device pairing schemes for both single and multi-antenna BSs as well as for multi-antenna devices. Moreover, an interference cancellation technique for asynchronous uplink NOMA systems is introduced in [72]. Furthermore, there are some studies that targeted to maximize the energy efficiency of the MTC with NOMA [73, 74].

In [75], a distributed power control algorithm is designed for the uplink of a NOMA system consisting of two cells. It is assumed that each cell has one BS and two devices, where the aim of each BS is to minimize the total power of its two devices, while their rate requirements are satisfied. This problem is formulated as a two-player non-cooperative game, where the BSs correspond to the players. The properties of the Nash equilibrium are investigated and a distributed algorithm is proposed that converges to it. The work in [76], considers a multi-cell network where to enhance the performance of cell-edge devices, fractional frequency reuse (FFR) is used. In the FFR, the entire bandwidth of the system is partitioned into a cell-center bands set and a cell-edge bands set. In order to avoid interference between neighboring cells, the cell-edge bands are split between the three neighboring cells. Furthermore, the cell devices are divided into cell-interior and cell-edge devices based on their receiving power from the serving BS. On the cell-center bands, only cell-interior devices are allowed to transmit using NOMA, while for the cell-edge bands, cell-interior and cell-edge devices can be scheduled together, as the large difference in channel gain between cell-interior and cell-edge devices is desirable in NOMA. In order to schedule devices for the uplink NOMA, a proportional fairness-based scheme is proposed.

2.3.2.2 Grant-free NOMA

Different form the aforementioned studies which use grant-based NOMA, the works in [77–80] exploit NOMA for the random access in multi-channel networks. More specifically, in [77] a set of power levels is defined, and each device with a packet for transmission randomly chooses a power level from that set and a channel. For this scheme, a closed-form expression for the lower bound of the throughput is derived. Furthermore, as the main drawback of this scheme is that the transmission power can be high, a channel-dependent selection scheme for the channel and power level is proposed which can reduce the transmission power. In addition, an upper bound for the average transmission power is obtained.

The above work is extended in [81], where device channels experience Rayleigh fading and pathloss. In particular, compared to [81], an improved lower bound for the throughput of this NOMA random access is derived. Furthermore, the average access delay is obtained, and the energy efficiency and the optimal retransmission probability are examined.

The drawback of the aforementioned scheme is that the set of power levels is predefined, thus it may lead to low performance for the network if the power levels are not suited. This issue is addressed in [79], where an adaptive set of power levels is used. In that work, to obtain the proper power level set, the number of active devices is estimated by the gateway and then based on that, the set is determined. In particular, a flexible frame structure is proposed which consists of five phases, shown in Fig. 2.8. In the first phase, a beacon is transmitted by the gateway to announce its readiness to receive packets. Then, in the second phase, devices having a packet transmit a training sequence to assist the gateway in detecting the number of active devices. The gateway estimates the number of active devices by performing a multiple hypotheses testing. It also adjusts its SIC receiver degree for the optimum power levels. It should be noted that if devices are registered with the gateway, there would be no need for using multi hypothesis testing but this will significantly increase the control phase length. In the third phase, if the estimated number of active devices is not in the range of the optimal power levels, the gateway aborts the transmission and starts the frame again by sending a new beacon which implies that the active devices use a random backoff. However, if the number of devices is in the range, the gateway broadcasts the SIC degree to the devices and each device chooses

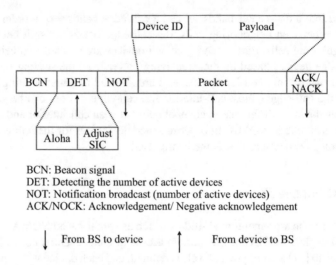

BCN: Beacon signal
DET: Detecting the number of active devices
NOT: Notification broadcast (number of active devices)
ACK/NOCK: Acknowledgement/ Negative acknowledgement

↓   From BS to device   ↑   From device to BS

Fig. 2.8 Proposed frame structure in [79]

one optimal power level randomly. If devices choose distinct power levels, then the gateway can decode the signal of each device and it sends an ACK. However, if distinct power levels are not chosen, the reselection process is continued. After a few attempts, if there is no successful transmission, the devices receive a NACK and enter a random back-off mode. In a similar approach, in [78], the estimated number of active devices is broadcast to the devices, and each device adjusts its transmission power accordingly.

In [80], the cell area is divided into different layers, based on predetermined inter-layer received power difference, shown in Fig. 2.9. More specifically, in this approach, each device having a packet for transmission, chooses its transmit power equal to its target received power divided by its channel gain. Furthermore, in order to effectively control the number of contenders and distribute the traffic over time, the EAB mechanism is used, where the parameter is broadcast by the BS to the devices. In the EAB mechanism, if a device wants to transmit, it generates a random number between 0 and 1. If the number is less than the EAB parameter, the device proceeds to transmit, otherwise it has to backoff temporarily.

In particular, the proposed access scheme consists of 5 following phases.

- Step 1: At first, the BS performs an optimization problem to obtain the EAB access control parameter, as well as the number of NOMA layers in each time-slot, and then it broadcasts this information.
- Step 2: At the device, if it has a packet in its queue, it generates a random number. If the random number is less than the EAB parameter, the device computes its transmission power based on its location, CSI and number of layers, and then it goes to step 3. Otherwise, it waits for the next available time-slot.
- Step 3: Each device transmits its packet with the calculated transmission power.

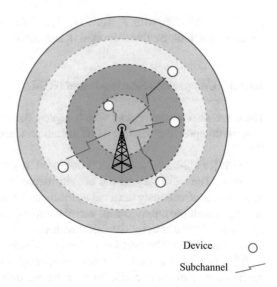

Fig. 2.9 Distributed layered grant-free non-orthogonal multiple access for massive MTC [80]

Device ○

Subchannel

- Step 4: If the device receives the ACK, it indicates that the transmission was successful. Otherwise it tries in next available time slot.

In [82], the performance of a p-persistent slotted ALOHA system in support of NOMA transmissions is investigated. More specifically, in this system, at each time-slot, with probability of p, the device transmits and with probability of $1 - p$, does not transmit. In order to benefit from NOMA, each device chooses its transmission power such that the received signal at the BS is v_1 or v_2 ($v_1 > v_2$), with different probabilities. The probability to choose v_1 is denoted by τ_1, and the probability to choose v_2 is represented by τ_2, where $\tau_1 + \tau_2 = p$. In order to maximize the achieved long-term average throughput of the system, the values of τ_1 and τ_2 are tuned by formulating the problem as a combinatorial optimization problem. To solve the problem, an iterative algorithm is proposed in which at each iteration, first τ_2 is updated to maximize the throughput, then with updated τ_2, τ_1 is updated to maximize the throughput. The iteration continues until the throughput improvement becomes negligible.

In [83], a distributed access scheme is proposed in which devices adaptively adjust their transmit power and control the transmission probability over time by observing the channel outcomes such as idle, collision and success. In particular, two predetermined received powers are considered, P_1 and P_2, where $P_1 > P_2$. In addition, devices are divided into two groups according to their distances from the BS. Devices located near the BS are denoted by N_1, while the rest are denoted by N_2. Devices that belong to the second group always transmit their packet with received power P_2 due to their low channel gain. On the other hand, devices of the first group choose either P_1 or P_2. At the end of each time-slot, the BS broadcasts the packet transmissions outcomes which can be one of the five cases as follows: one successful packet transmission with P_i for $i = 1, 2$, idle, two successful transmissions, or collision. Devices use this information and decide proper actions.

2.3.2.3 Compressed Sensing-Based NOMA

There are also several works which deploy code-domain NOMA [84–87], where various compressive sensing (CS) techniques are used for multi-user detection (MUD).

In [88], a CS-based approach is proposed to jointly estimate the device activities and data. The reliability of the activity detection is of high importance, since if the device is erroneously detected as inactive, the data is lost. To estimate the device activity, a multiple measurement vector compressed sensing approach is used, which allows for device activity detection with complexity invariant of the length of the transmitted frame. The detail of the proposed scheme is as follows.

It is assumed that a set of D devices sporadically access the channel to transmit modulation symbols to a BS. At each frame, only a set of devices are active and transmit data with length L over the whole frame. Data of these devices is denoted by X, which is a $D \times L$ matrix. Furthermore, it is assumed that each device uses a specific random spreading sequences of length m for multi user detection. These sequences are represented by T, which is a $m \times D$ matrix. Consequently, a per-frame detection model is as follows

$$Y = TX + W, \tag{2.3}$$

where W represents i.i.d. samples from a white Gaussian noise process. In the proposed scheme, the focus is to estimate device activity without estimating the underlying data. Data detection can be done after the set of active devices has been estimated correctly. In order to perform device activity detection, the covariance matrix of the received signal is considered. However, as the covariance matrix is not available at the detector, the sample covariance is used, which is calculated as

$$\phi_{YY} = \frac{1}{L} YY^H = TGT^H + \phi_{WW}, \tag{2.4}$$

where $G = Q_{XX}$ is a diagonal matrix whose $g_{d,d}$-th entry is one if the d-th device is active and zero otherwise. In order to find the diagonal elements of G, two approaches are proposed. First, a matrix matching pursuit (MMP) algorithm is developed which is an extension of the orthogonal matching pursuit (OMP) for solving underdetermined matrix problems. In contrast to the OMP, the MMP does not require a matrix inversion. Furthermore, an approximate maximum a posteriori probability MAP detection scheme is proposed that refers to the solution of a regularized least squares problem.

The performance of the MMV approach mainly depends on the length of spreading sequences, the number of devices, the sparsity, and the background noise. In general, the ratio of the length of spreading sequences to the number of devices is a key parameter that determines the number of the active devices that can be successfully detected. Therefore, in order to improve the performance, longer spreading codes or a wider system bandwidth is required.

In [89], a power-domain NOMA is applied to improve the performance of CS-based schemes for a given length of spreading codes or a system bandwidth with multiple layers in the power domain. In particular, it is assumed that there are Q different layers, where each layer is characterized by a different received signal at the BS. Devices are evenly divided over the layers, therefore the number of devices at each layer, M, is D/Q, where D denotes the number of devices. An illustration of transmitted signals for random access in power and code domains is shown in Fig. 2.10. In each layer, there are M spreading codes which are not orthogonal to each other. In order to detect the signals of each layer, SIC operations is performed. The success of SIC depends on the background noise and interference from other layers. Therefore, to perform the SIC with a high success probability, the proper power level for each layer is derived.

In [90], a more realistic scenario is considered in which although devices transmit their packets sporadically, however some of them with a high probability transmit their data in adjacent time-slots. In other words, the active device set changes over time but it changes slowly. Therefore, the temporal correlation of active device sets in several continuous time-slots can be exploited to enhance the multi user detection performance. In [90], a dynamic compressive sensing-based approach is proposed which takes into account the temporal correlation between continuous time-slots. The main idea is that instead of initiating with an empty set, the proposed algorithm starts with the previously estimated device set.

Although, the above works enhance the performance of considered networks, however none of them address the unique requirements of an M2M network, which consists of a large number of devices with heterogeneous traffic rates. In these networks, using only grant-based NOMA schemes lead to poor performance for devices with sporadic transmissions while NOMA-based random access may not perform well due to a large number of devices. In fact, the chance of choosing the same power and channel increases with increasing number of devices, leading to a large collision rate. Furthermore, CS-based schemes are only applicable for the scenarios that device activity is time-related and sporadic [91].

2.4 Massive MIMO for Massive MTC

Another promising techniques to provide massive connectivity in massive MTC use cases is massive multiple-input multiple-output (MIMO). This technique employs large number of antennas at the BS to fulfill the demand for massive access. In particular, the large number of antennas creates large number of spatial degrees of freedom leading to the remarkable properties including channel hardening and favorable propagation. The occurrence of channel hardening essentially means that

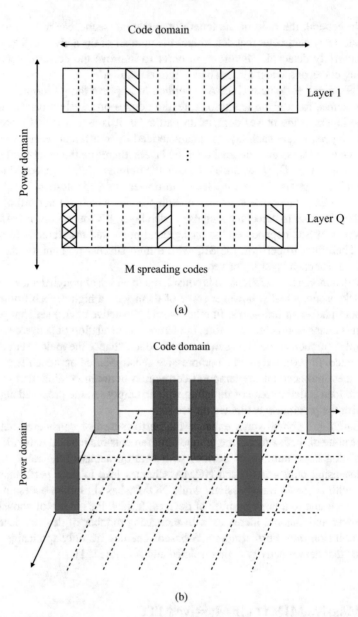

Fig. 2.10 NOMA-based compressive random access using Gaussian spreading in power and code domains [89]. (**a**) 2-dimensional multiple access scheme. (**b**) An illustration of transmitted signals for random access in the power and code domains

the massive multi-antenna prior post-processing transforms the channel into almost deterministic quantities, depending only on large scale fading parameters. The occurrence of favorable propagation means that as the size of the antenna array becomes large, the channel vectors of device become asymptotically orthogonal. Therefore, the BS equipped with massive number of antennas becomes more efficient in mitigating inter-user interference through the use of spatial multiplexing techniques [92–95].

In the following, access schemes proposed for MTC systems, in which the BS is equipped with a massive number of antennas, are presented.

2.4.1   Grant-Based Random Access Schemes

One of the grant-based random access schemes proposed for massive MIMO systems is strongest-user collision resolution (SUCRe), which is described as follows.

In SUCRe, devices having a packet for transmission, randomly choose a preamble from an orthogonal set of preambles. As the number of preambles is lower than the number of devices, some devices may choose the same preamble resulting in a preamble collision [96]. Channel estimations are done on the received preambles, consequently for colliding devices, channel estimates are contaminated. The estimates are used by the BS to perform precoded transmission. At the device side, the received signal array gain is measured. If the array gain is equal to the number of antennas, it means that no collision has happened and the device is granted access to the preamble sequence. However, if the array gain is a fraction of the number of antennas, it indicates that collision has occurred in the preamble domain. In order to resolve the collision, a distributed decision rule is employed at devices, such that only the strongest user is granted access to the preamble sequence. It has been shown that SUCRe is able to resolve 90% of collisions and it is highly scalable in terms of number of devices as it uses a distributed algorithm.

In order to further enhance the performance of SUCRe, variants of this scheme have been proposed in the literature. For example, in [97], the preambles that are not chosen by any devices are assigned to the collided devices that lost in the collision resolution phase of the scheme. However, this approach requires additional signaling.

Furthermore, in the classical SUCRe protocol, to perform retransmissions, a hard decision rule is employed. For crowded scenarios, the performance of this scheme may not be satisfactory, as the number of collisions with a higher number of contending devices increases. Therefore, the idea of using a soft decision retransmission rule is proposed in [98], where the retransmission probability leads to improvement to the SUCRe in crowded scenarios.

In addition, SUCRe often privileges devices having the highest signal, consequently the scheme is unfair for devices that are far from the BS. To address this issue, in [99], an access scheme is proposed which deploys access class barring with power control. The proposed scheme allows each device to estimate the number of collided devices for the chosen preamble and obtain the ACB factor to determine the preamble retransmission probability in the next protocol step.

2.4.2   Grant-free Random Access Schemes

Different from grant-based random access schemes, in grant-free random access schemes, both preamble sequence and data are transmitted as part of the initial access attempt. In these schemes, no collision resolution is performed, therefore they are simplified. However, in order to properly decode the signal of transmitting devices, improved transmission strategies or decoding strategies are needed.

In some protocols, the transmission of the devices accessing the network is organized in multiple slots. For example, in [100], time is divided into frames consisting of multiple time-slots as illustrated in Fig. 2.11. In the proposed scheme, when a device has a codeword to transmit, it splits the codeword into multiple parts and transmits the codeword parts in multiple time-slots. Furthermore, each device is assigned a unique predefined pseudo-random pilot hopping pattern, where pilot sequences denoted by Q_i are orthogonal. To transmit each codeword part, the device selects the pilot according to this pattern. As the BS knows the pilot-hopping patterns of all devices, by using a correlation decoder across the time-slots, the activated pilot-hopping patterns can be detected.

The advantage of using this approach is that with coding spreading across a large number of transmission time-slots, the interference is averaged out. Therefore, from an information-theoretic point of view, it is possible to define a reliable transmission rate.

(2) Coded RA: An alternative approach is proposed in [101], in which time is divided into frames and each frame consists of multiple time-slots as illustrated in Fig. 2.12. At each time-slot, with a certain activation probability, a device chooses a random preamble and transmits the preamble followed by the data over the time-slot. The same data is retransmitted in each frame. In order to decode the data, if a collision-free transmission happens at the given time-slot, the BS is able to decode the data of corresponding device successfully. Furthermore, using successive interference cancellation technique, the contribution of the decoded

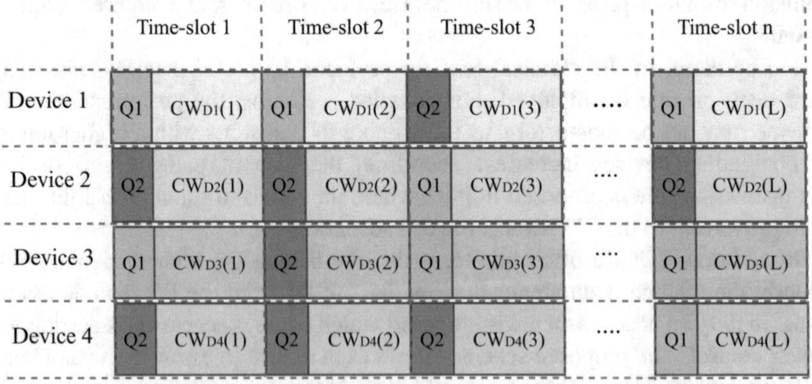

	Time-slot 1		Time-slot 2		Time-slot 3			Time-slot n	
Device 1	Q1	$CW_{D1}(1)$	Q1	$CW_{D1}(2)$	Q2	$CW_{D1}(3)$	....	Q1	$CW_{D1}(L)$
Device 2	Q2	$CW_{D2}(1)$	Q2	$CW_{D2}(2)$	Q1	$CW_{D2}(3)$	....	Q2	$CW_{D2}(L)$
Device 3	Q1	$CW_{D3}(1)$	Q2	$CW_{D3}(2)$	Q1	$CW_{D3}(3)$	....	Q1	$CW_{D3}(L)$
Device 4	Q2	$CW_{D4}(1)$	Q2	$CW_{D4}(2)$	Q2	$CW_{D4}(3)$	....	Q1	$CW_{D4}(L)$

Fig. 2.11 Random pilot and data access in massive MIMO systems [100]

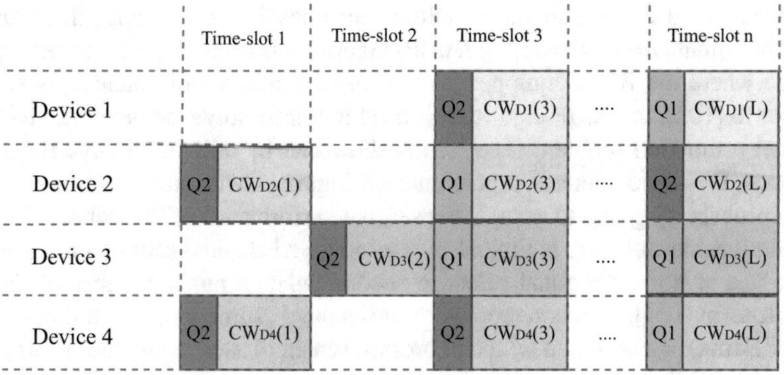

	Time-slot 1	Time-slot 2	Time-slot 3		Time-slot n
Device 1			Q2 $CW_{D1}(3)$	....	Q1 $CW_{D1}(L)$
Device 2	Q2 $CW_{D2}(1)$		Q1 $CW_{D2}(3)$	....	Q2 $CW_{D2}(L)$
Device 3		Q2 $CW_{D3}(2)$	Q1 $CW_{D3}(3)$	....	Q1 $CW_{D3}(L)$
Device 4	Q2 $CW_{D4}(1)$		Q2 $CW_{D4}(3)$	....	Q1 $CW_{D4}(L)$

Fig. 2.12 Coded pilot random access for massive MIMO systems [101]

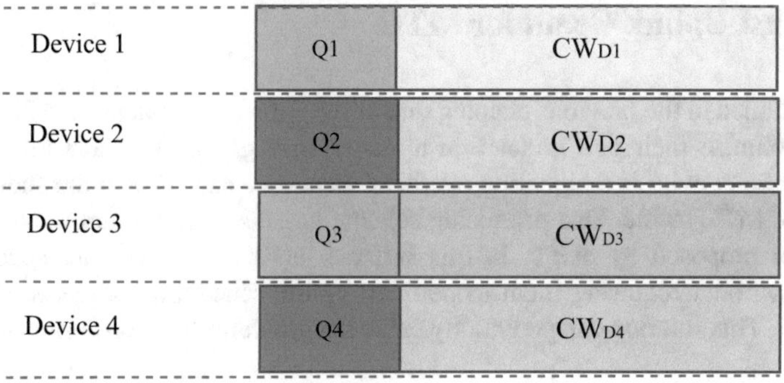

Device 1	Q1	CW_{D1}
Device 2	Q2	CW_{D2}
Device 3	Q3	CW_{D3}
Device 4	Q4	CW_{D4}

Fig. 2.13 Grant-free compressed sensing-based access scheme for massive MIMO systems [102]

packet is deducted in the previous time slots thus possibly removing a transmission that was causing a collision.

2.4.3   Grant-free Access Relying on Compressed Sensing

Another promising solution to detect device activity in mMTC is to utilize compressed sensing-based approach (Fig. 2.13), due to the sporadic nature of the MTC device activities. In the following, we review some of the schemes in which each device is assigned a unique preamble sequence. However, as the number of orthogonal sequences are limited by the channel coherence time, pseudo-orthogonal sequences are used. Due to the non-orthogonality of the preamble sequences, the channel estimates cannot be easily obtained by correlating the received signal with each preamble sequence.

In [102], the joint device activity detection and channel estimation problem are formulated as a compressed sensing problem, considering scenarios the BS is equipped with a single antenna or multiple antennas. In order to solve this problem, the approximate message passing (AMP) algorithm is used. In [103], an asymptotic regime where the BS is equipped with a massive number of antennas is studied. Similar to [102], the AMP algorithm is used to jointly solve the device activity and channel estimation problem. The results show that by utilizing a large number of antennas, the device detection performance improves with misdetection and false alarm probability going to zero. However, the performance of this scheme in terms of overall achievable rate is limited by the increased channel estimation errors, due to the use of non-orthogonal pilots to accommodate a larger number of devices. Therefore, in [104], to obtain more accurate channel estimation, the minimum-mean square estimator, combined with compressed sensing-based techniques is proposed.

2.5 Fast Uplink Grant for MTC

As mentioned in the previous chapter, one of the main requirements in MTC system is to maximize their performance in terms of throughput. To reach this goal, an access scheme with low signaling overhead should be used due to the short-packet nature of MTC traffic. One promising scheme is a fast uplink grant method which has been proposed by 3GPP. In this scheme, uplink resources are accessed by devices without requiring them to perform uplink scheduling requests, shown in Fig. 2.14. This solution can potentially solve the problems that conventional random

Fig. 2.14 Comparison of fast grant uplink and RA procedure

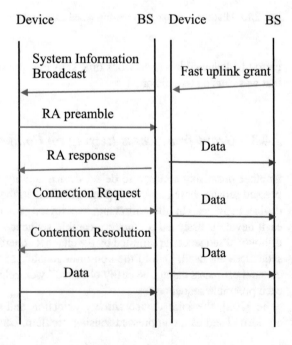

access schemes face in MTC such as collisions, congestion, and inefficiency due to large signaling overhead compared to the actual data size. Naturally, device selection for fast uplink grant and uplink radio resource allocation are performed at the BS [105].

In order to adopt the fast uplink grant scheme, sophisticated source traffic prediction mechanisms are needed to predict which devices are active at each time. Otherwise, without proper predictions, uplink grant allocation can lead to a waste of resources. In the literature, most of the works have focused on modeling and analyzing the aggregated traffic at the BS or network for optimizing the operation of the network. However the traffic prediction problem at user/device level is only addressed in a few works in which independent traffic models for devices are considered. While, in MTC, devices are likely to produce more correlated and predictable traffic patterns, e.g., if the traffic is generated based on observations of some common physical phenomenon.

2.5.1 Existing Works

As discussed in the above, the traffic prediction problem can be classified into two groups: (1) aggregated traffic at the network, (2) device traffic.

In the following, first, we review some existing works at the cellular level and then we study a few works that used traffic prediction at the source level.

2.5.1.1 Cellular Traffic Prediction

In [106], a long short-term memory (LSTM)-based model is used to predict the traffic of a BS in the next time-slot, where the proposed architecture is shown in Fig. 2.15. In order to increase the accuracy of the prediction, a multilayer structure is proposed, where each layer consists of multiple LSTM units to learn the temporal dependencies of the traffic. In fact, the number of concatenated units indicates the number of observations of the data that are considered before making the prediction.

In [107], the traffic pattern of 9000 cellular BSs is analyzed. To model the traffic, a time series approach is applied in which the traffic is decomposed into two components: regular component and unpredictable random component. In order to forecast the regular component, a seasonal autoregressive moving average (SARIMA) model is used. Furthermore, to predict the stochastic component, cross correlation between the stochastic component of a cellular BS and its neighbouring BSs is computed. The results show that the average cross correlation value is low indicating that stochastic components of neighbouring cellular BSs are nearly unrelated.

In [108], to forecast the traffic of BSs, a multi-task learning architecture is proposed. This architecture consists of a shared learning machine and a dedicated learning machine for each BS, shown in Fig. 2.16. The intuition behind this

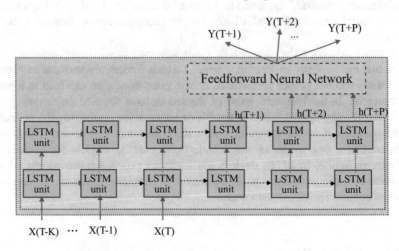

Fig. 2.15 LSTM-based architecture for the cellular traffic prediction [106]

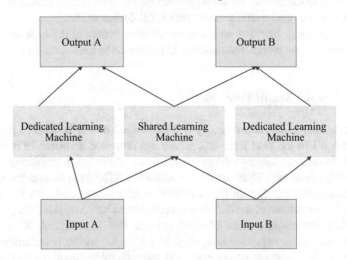

Fig. 2.16 RNN-based architecture for the cellular traffic prediction [108]

architecture is that the dedicated learning machine captures the specific features of the corresponding BS while the shared learning machine can capture the spatial dependency among the BSs. The output of these machines are passed to a fully connected feedforeword neural network to predict the traffic of the corresponding BS.

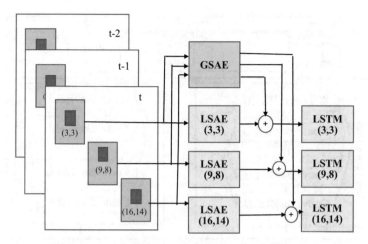

Fig. 2.17 Autoencoder-based architecture for the cellular traffic prediction [109]

In [109], both temporal and spatial dependencies of BSs are taken into account to model and predict the cellular traffic. A hybrid deep learning model is presented consisting of an autoencoder-based deep model for spatial modeling and LSTM units for temporal modeling. The proposed model is shown in Fig. 2.17, where the red rectangle represents the target BS and the blue rectangle represents its neighboring BSs. Each data patch consists of the data of the target BSs and its neighbors, which is passed to the corresponding local stacked autoencoder (LSAE). Furthermore, a global stacked autoencoder (GSAE) is considered in the model, which takes the data patch of all BSs as input and produces an encoded representation. After encoding by the GSAE and LSAE, each data patch, represented by the concatenated output of these two encoders, is passed to the corresponding LSTM for prediction.

In fact, in this model, the GSAE captures the common characteristics among BSs while the LSAE captures the specific features of each BS. The advantage of using this model is that since only the GSAE takes all the data patches as input, and LSAEs only take their corresponding data patch, it is faster to train the model. Furthermore, LSAEs can be trained in parallel as they are independent from each other.

In [110], temporal and spatial dependence of traffic among different cells are investigated. More specifically, the traffic data is treated as an image and a convolutional neural network (CNN)-based approach is used to jointly capture the temporal and spatial dependency among cells. The city is split into a grid with size of $H \times W$ and each square of the grid is referred to as a cell. In order to predict the traffic at time slot t, the intervals before t are divided into two segments, i.e., recent time and daily history. The traffic of recent time segment models the temporal closeness dependence while the traffic sampled from daily history models the temporal period dependence. Data of each segment is passed to a convolutional learning system and to capture the relationship between the output of these systems,

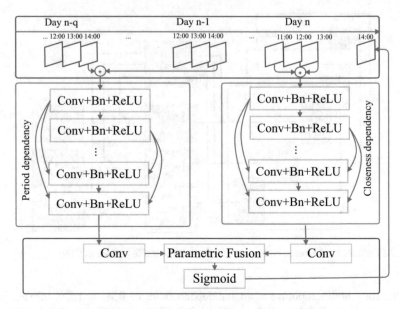

Fig. 2.18 CNN-based architecture for the cellular traffic prediction [110]

a parametric matrix-based scheme is proposed fusing the features of closeness and period. The proposed architecture is shown in Fig. 2.18.

In [111], to model both the temporal and spatial aspects of the BSs traffic, a graph convolutional (GC) network embedded LSTM model is used. More specifically, to capture the spatial dependency among BSs the spatial distance between them is used. That is the spatial dependence of two BSs decreases by increasing their spatial distance. Hence, to model the spatial dependency among BSs, a dependency graph is defined in which each BS is represented by a vertex and there is an edge between two vertices if their spatial distance is less than a certain threshold. In addition, not only the recent demand history is applied to forecast the future demands, but also the periodic history, e.g., day(s) ahead demands, is considered in order to obtain an accurate demand predictor.

In [111], temporal and spatial correlation between traffic demands of cells are considered into account to forecast the traffic of each cell. More specifically, the problem is formulated as a GC-LSTM. The proposed architecture is shown in Fig. 2.19.

2.5.1.2 Device Traffic Prediction

As mentioned earlier, source/device traffic prediction is only studied in a few works. In the following, we discuss the approaches that are so far proposed to predict the

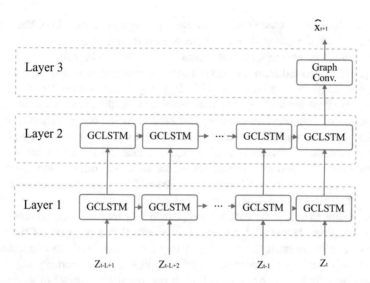

Fig. 2.19 GC-LSTM-based architecture for the cellular traffic prediction [111]

traffic of each device. These predictions are generally possible in MTC systems as most of devices are stationary or have low mobility. Furthermore, number of devices that are associated to a BS is often fixed [112].

Traffic prediction at device level can be classified into two groups: periodic reporting and event-driven transmissions. In periodic reporting, devices transmit their packets periodically at specific times, while in event-driven traffic, a number of devices transmit their requests to report a certain event. It is obvious that the periodic traffic prediction is easier than the event-driven traffic. In the following, prediction algorithms for both type of traffics are presented.

Periodic traffic prediction: Many monitoring applications are dependent on machine-type devices to periodically transmit sensory data sampled from the physical environment. However, these applications generate packets in different time intervals. Therefore, the BS must learn exact time instances at which any machine-type device generates its packet. Assuming that the BS has historical transmission data of devices, machine learning algorithms can be used to predict the traffic of each device. However, the prediction must be accurate, since some applications have strict latency requirements less than 10 ms.

Event-driven traffic prediction: In some machine-type applications, when a specific event occurs, several devices that detect the event should report it to the BS. This can cause a burst of RA scheduling requests, which cannot be efficiently handled. On the other hand, it is not possible to predict the event that has not observed yet, but the BS can use the historical data of previous events, to predict which devices face the same event. For example, in [113], a predictive approach for resource allocation is used in which an event propagates along a line. Here, the BS can learn which devices will initiate transmission in case of an event. However, this method cannot be generalized to all possible types of MTC events.

In [114], a directed information (DI) learning framework is proposed to predict the source traffic in event-driven MTC. In [114], it is assumed that the network consists of static devices or with low mobilities. In the proposed scheme, for a long period of time, devices use the grant-based random access scheme, in which first they transmit a scheduling request and then if a resource is allocated to them, they transmit their packets. It is also assumed that the BS keeps the history of transmissions. For periodic transmissions, the historical information can be used to learn the time instants that devices have packets for transmission. Therefore, after learning the pattern of periodic transmissions, devices no longer use the random access scheme to access the channel and instead they transmit their packet by using fast uplink grant. However, for unexpected events that happen out of the pattern, devices have to transmit the scheduling requests which leads to congestion and consequently waste of resources. To address this issue, a learning algorithm is proposed in which upon event detection, it can predict which set of devices have a packet for transmission. These devices with high probabilities of having a packet for transmission are allocated time-slots and consequently the congestion can be avoided and higher throughput can be achieved. The drawback of this scheme is that the directed information should be calculated for each pair of devices, therefore it may suffer from scalability.

The authors of [115] have only considered the temporal correlation to predict the M2M traffic. More specifically, a time series framework is proposed to model both the asynchronous and synchronous M2M device arrival processes over a large time scale and forecast the upcoming M2M network traffic dynamics at prospectively sequential time points.

2.6   Concluding Remarks

Although the discussed schemes aimed at maximizing the throughput of M2M networks, the scheduling-based and hybrid schemes still suffer from large over-heads, while the random access schemes cannot achieve optimality due to collisions. In fact, in these networks, to achieve the maximum spectral efficiency, taking into account the device traffic statistics can play an important role. Using this information, the AP is able to allocate resources to devices in a proactive manner, which reduces the signaling overhead required in the request-based scheduling schemes. Furthermore, as device traffic statistics are considered in the scheduling procedure, the probability that the allocated resources left idle is minimized. Moreover, for scenarios consisting of heterogeneous devices, reconfigurable access

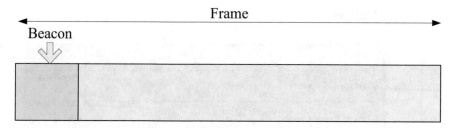

Fig. 2.20 General structure of the proposed frame

schemes can cope with traffic characteristics of different device categories, i.e., periodic or sporadic type traffics.

The aforementioned benefits of using traffic-aware and reconfigurability in access schemes motivated us to design schemes with these characteristics for MTC systems. In particular, we aim to maximize the throughput of these networks while the application requirements are met. To that end, we develop access schemes working in a frame-by-frame basis, where the general proposed frame structure is shown in Fig. 2.20. More specifically, in Chap. 3, we propose an access scheme in which the whole frame is allocated to a CSMA scheme with deterministic backoffs. In Chaps. 4 and 5, the frame is divided into two segments: demand-free assignment and random-based access. In Chap. 6, we exploit the same frame structure as in Chaps. 4 and 5, where devices in each segments are exclusive. In particular, one segment belongs to LTE devices while the other one is allocated to WiFi devices. In Chap. 7, the frame structure is further split into three segments, where an additional segment belongs to the NOMA demand-free assignment-based schemes. Finally, in Chap. 8, we consider the frame structure proposed in Chap. 4, however instead of using the demand-free assignment access scheme, we deploy a self-organized TDMA scheme, in which devices reserve a time-slot in a distributed manner.

Although the proposed frame structure is represented in one dimension, it can be applied for multi-frequency systems as well. The equivalent frame structure for multi-carrier systems is shown in Fig. 2.21. In this structure, radio resources are divided into two dimensions; frequency domain and time domain, where the portion of resources allocated to each segment is the same as the 1-dimension representation. As an example, Fig. 2.22 represents the equivalent for the frame structure proposed in Chap. 7.

In conclusion, in this chapter, we have provided a comprehensive survey on the existing MTC access schemes and discussed their drawbacks in meeting the requirements of M2M networks. To address these shortcomings, the idea of using traffic-aware and reconfigurable access schemes has been proposed in this book, where the further details of these schemes are discussed in subsequent chapters.

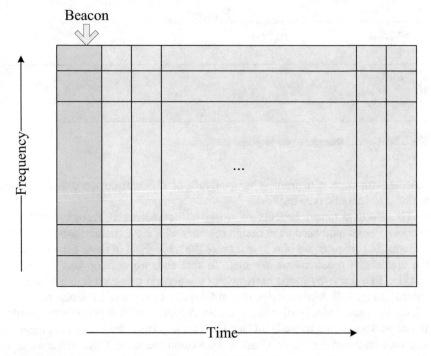

Fig. 2.21 General structure of the proposed frame in 2 dimensions

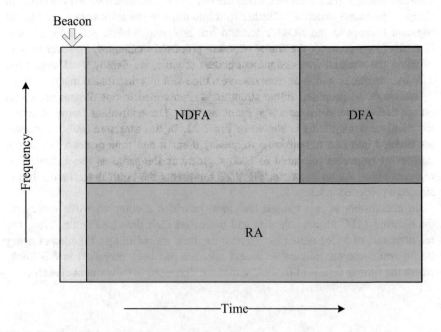

Fig. 2.22 General structure of the proposed frame in 2 dimensions with 3 segments

References

1. A.E. Mostafa, Y. Gadallah, A statistical priority-based scheduling metric for M2M communications in LTE networks. IEEE Access **5**, 8106–8117 (2017)
2. E. Soltanmohammadi, K. Ghavami, M. Naraghi-Pour, A survey of traffic issues in machine-to-machine communications over LTE. IEEE Internet Things J. **3**(6), 865–884 (2016)
3. A.G. Gotsis, A.S. Lioumpas, A. Alexiou, M2M scheduling over LTE: challenges and new perspectives. IEEE Veh. Tech. Mag. **7**(3), 34–39 (2012)
4. N. Afrin, J. Brown, J.Y. Khan, A delay sensitive LTE uplink packet scheduler for M2M traffic, in *IEEE Globecom Workshops (GC Wkshps), Atlanta, GA, USA* (2013)
5. P. Si, J. Yang, S. Chen, H. Xi, Adaptive massive access management for QoS guarantees in M2M communications. IEEE Trans. Veh. Technol. **64**(7), 3152–3166 (2015)
6. A. Elhamy, Y. Gadallah, BAT: a balanced alternating technique for M2M uplink scheduling over LTE, in *Proceedings of the IEEE Vehicular Technology Conference (VTC), Glasgow, UK* (2015)
7. A.M. Maia, D. Vieira, M.F. de Castro, Y. Ghamri-Doudane, A mechanism for uplink packet scheduler in LTE network in the context of machine-to-machine communication, in *Proceedings of the IEEE Global Communication Conference (GLOBECOM), Austin, TX, USA* (2014), pp. 2776–2782
8. M. Hasan, E. Hossain, D. Niyato, Random access for machine-to-machine communication in LTE-advanced networks: issues and approaches. IEEE Commun. Mag. **51**(6), 86–93 (2013)
9. S.-Y. Lien, J. Lee, Y.-C. Liang, Random access or scheduling: optimumLTE licensed-assisted access to unlicensed spectrum. IEEE Commun. Lett. **20**(3), 590–593 (2016)
10. L. Tello-Oquendo, V. Pla, I. Leyva-Mayorga, J. Martinez-Bauset, V. Casares-Giner, L. Guijarro, Efficient random access channel evaluation and load estimation in LTE-A with massive MTC. IEEE Trans. Veh. Technol. **68**(2), 1998–2002 (2018)
11. S.K. Sharma, X. Wang, Towards massive machine type communications in ultra-dense cellular IoT networks: current issues and machine learning-assisted solutions. IEEE Commun. Surv. Tutorials (2019)
12. H. He, Q. Du, H. Song, W. Li, Y. Wang, P. Ren, Traffic-aware ACB scheme for massive access in machine-to-machine networks, in *Proceedings of the IEEE International Conference Communication (ICC), London, UK* (2015)
13. H. Jin, W.T. Toor, B.C. Jung, J.-B. Seo, Recursive pseudo-bayesian access class barring for M2M communications in LTE systems. IEEE Trans. Veh. Technol. **66**(9), 8595–8599 (2017)
14. S. Duan, V. Shah-Mansouri, Z. Wang, V.W. Wong, D-ACB: adaptive congestion control algorithm for bursty M2M traffic in LTE networks. IEEE Trans. Veh. Technol. **65**(12), 9847–9861 (2016)
15. 3GPP TR 37.868 V11. 0.0, in *Study on RAN Improvements for Machine-type Communications* (2011)
16. R.-G. Cheng, J. Chen, D.-W. Chen, C.-H. Wei, Modeling and analysis of an extended access barring algorithm for machine-type communications in LTE-A networks. IEEE Trans. Wireless Commun. **14**(6), 2956–2968 (2015)
17. N. Zangar, S. Gharbi, M. Abdennebi, Service differentiation strategy based on MACB factor for M2M communications in LTE-A networks, in *Proceedings of the IEEE Consumer Communication and Networking Conference (CCNC), Las Vegas, NV, USA* (2016)
18. T.-M. Lin, C.-H. Lee, J.-P. Cheng, W.-T. Chen, PRADA: prioritized random access with dynamic access barring for MTC in 3GPP LTE-A networks. IEEE Trans. Veh. Technol. **63**(5), 2467–2472 (2014)
19. M. Vilgelm, H.M. Gürsu, W. Kellerer, M. Reisslein, LATMAPA: load-adaptive throughput-maximizing preamble allocation for prioritization in 5G random access. IEEE Access **5**, 1103–1116 (2017)
20. E.U.T.R. Access, *Study on RAN Improvements for Machine-type Communications*, TR 37.868 V. 11.0. 0. Technical Report

21. M. Shirvanimoghaddam, Y. Li, M. Dohler, B. Vucetic, S. Feng, Probabilistic rateless multiple access for machine-to-machine communication. IEEE Trans. Wireless Commun. **14**(12), 6815–6826 (2015)
22. M.S. Ali, E. Hossain, D.I. Kim, LTE/LTE-A random access for massive machine-type communications in smart cities. IEEE Commun. Mag. **55**(1), 76–83 (2017)
23. T. Adame, A. Bel, B. Bellalta, J. Barcelo, M. Oliver, IEEE 802.11 AH: the WiFi approach for M2M communications. IEEE Wireless Commun. **21**(6), 144–152 (2014)
24. L. Zheng, M. Ni, L. Cai, J. Pan, C. Ghosh, K. Doppler, Performance analysis of group-synchronized DCF for dense IEEE 802.11 networks. IEEE Trans. Wireless Commun. **13**(11), 6180–6192 (2014)
25. N. Nawaz, M. Hafeez, S.A.R. Zaidi, D.C. McLernon, M. Ghogho, Throughput enhancement of restricted access window for uniform grouping scheme in IEEE 802.11 AH, in *Proceedings of the IEEE International Conference Communication (ICC), Paris, France* (2017)
26. C.W. Park, D. Hwang, T.-J. Lee, Enhancement of IEEE 802.11 ah MAC for M2M communications. IEEE Commun. Lett. **18**(7), 1151–1154 (2014)
27. Y. Yang, S. Roy, Grouping-based MAC protocols for EV charging data transmission in smart metering network. IEEE J. Sel. Areas Commun. **32**(7), 1328–1343 (2014)
28. T.-C. Chang, C.-H. Lin, K.C.-J. Lin, W.-T. Chen, Traffic-aware sensor grouping for IEEE 802.11 ah networks: regression based analysis and design. IEEE Trans. Mobile Comput. **18**(3), 674–687 (2019)
29. Y. He, X. Ma, Deterministic backoff: toward efficient polling for IEEE 802.11 e HCCA in wireless home networks. IEEE Trans. Mobile Comput. **10**(12), 1726–1740 (2011)
30. *IEEE Standard Part 11: wireless LAN Medium Access Control (MAC) and Physical Layer (PHY) Specifications* (2007)
31. Y. Liu, C. Yuen, J. Chen, X. Cao, A scalable hybrid MAC protocol for massive M2M networks, in *Proceedings of the IEEE Wireless Communication Network Conference (WCNC), Shanghai, China* (2013)
32. Y. Liu, C. Yuen, X. Cao, N.U. Hassan, J. Chen, Design of a scalable hybrid MAC protocol for heterogeneous M2M networks. IEEE Internet Things J. **1**(1), 99–111 (2014)
33. G.C. Madueño, Č. Stefanović, P. Popovski, Reliable and efficient access for alarm-initiated and regular M2M traffic in IEEE 802.11 ah systems. IEEE Internet Things J. **3**(5), 673–682 (2016)
34. A. Laya, C. Kalalas, F. Vazquez-Gallego, L. Alonso, J. Alonso-Zarate, Goodbye, Aloha!. IEEE Access **4**, 2029–2044 (2016)
35. A.-T.H. Bui, C.T. Nguyen, T.C. Thang, A.T. Pham, Design and performance analysis of a novel distributed queue access protocol for cellular-based massive M2M communications. IEEE Access **6**, 3008–3019 (2018)
36. F. Chaves, A. Cavalcante, E. Almeida, F. Abinader Jr, R. Vieira, S. Choudhury, K. Doppler, LTE/Wi-Fi coexistence: challenges and mechanisms, in *XXXI Simposio Brasileiro de Telecomunicacoes* (2013)
37. S. Sagari, I. Seskar, D. Raychaudhuri, Modeling the coexistence of LTE and WiFi heterogeneous networks in dense deployment scenarios, in *IEEE International Conference on Communication Workshop (ICCW), London, UK* (2015)
38. F.M. Abinader, E.P. Almeida, F.S. Chaves, A.M. Cavalcante, R.D. Vieira, R.C. Paiva, A.M. Sobrinho, S. Choudhury, E. Tuomaala, K. Doppler et al., Enabling the coexistence of LTE and Wi-Fi in unlicensed bands. IEEE Trans. Commun. **52**(11), 54–61 (2014)
39. B. Chen, J. Chen, Y. Gao, J. Zhang, Coexistence of LTE-LAA and Wi-Fi on 5 GHz with corresponding deployment scenarios: a survey. IEEE Commun. Surv. Tutorials **19**(1), 7–32 (2017)
40. A. Mukherjee, J.-F. Cheng, S. Falahati, H. Koorapaty, R. Karaki, L. Falconetti, D. Larsson et al., Licensed-assisted access LTE: coexistence with IEEE 802.11 and the evolution toward 5G. IEEE Trans. Commun. **54**(6), 50–57 (2016)
41. H. Cui, V.C. Leung, S. Li, X. Wang, LTE in the unlicensed band: overview, challenges, and opportunities. IEEE Wireless Commun. **24**(4), 99–105 (2017)

42. S. Hajmohammad, H. Elbiaze, Unlicensed spectrum splitting between femtocell and WiFi, in *Proceedings of the IEEE International Conference Communication (ICC), Budapest, Hungary* (2013)
43. 3GPP Study Item RP-141397, Study on licensed-assisted access using LTE. Technical Report (2014)
44. S. Dama, A. Kumar, K. Kuchi, Performance evaluation of LAA-LBT based LTE and WLAN's co-existence in unlicensed spectrum, in *Proceedings of the IEEE Global Communication Conference (GLOBECOM), San Diego, CA, USA* (2015)
45. H. Ko, J. Lee, S. Pack, A fair listen-before-talk algorithm for coexistence of LTE-U and WLAN. IEEE Trans. Veh. Technol. **65**(12), 10116–10120 (2016)
46. C. Cano, D.J. Leith, Coexistence of WiFi and LTE in unlicensed bands: a proportional fair allocation scheme, in *IEEE International Conference on Communication Workshop (ICCW), London, UK* (2015)
47. R. Yin, G. Yu, A. Maaref, G. Y. Li, LBT-based adaptive channel access for LTE-U systems. IEEE Trans. Wireless Commun. **15**(10), 6585–6597 (2016)
48. X. Wang, T.Q. Quek, M. Sheng, J. Li, Throughput and fairness analysis of Wi-Fi and LTE-U in unlicensed band. IEEE J. Sel. Areas Commun. **35**(1), 63–78 (2017)
49. Q. Zhang, Q. Wang, Z. Feng, T. Yang, Design and performance analysis of a fairness-based license-assisted access and resource scheduling scheme. IEEE J. Sel. Areas Commun. **34**(11), 2968–2980 (2016)
50. M. Mehrnoush, V. Sathya, S. Roy, M. Ghosh, Analytical modeling of Wi-Fi and LTE-LAA coexistence: throughput and impact of energy detection threshold. IEEE/ACM Trans. Netw. **26**(4), 1990–2003 (2018)
51. M. Mehrnoush, S. Roy, V. Sathya, M. Ghosh, On the fairness of Wi-Fi and LTE-LAA coexistence. IEEE Trans. Cognitive Commun. Netw. **4**(4), 735–748 (2018)
52. Y. Gao, S. Roy, Achieving proportional fairness for LTE-LAA and Wi-Fi coexistence in unlicensed spectrum. IEEE Trans. Wireless Commun. **19**(5), 3390–3404 (2020)
53. S. Han, Y.-C. Liang, Q. Chen, B.-H. Soong, Licensed-assisted access for LTE in unlicensed spectrum: a MAC protocol design. IEEE J. Sel. Areas Commun. **34**(10), 2550–2561 (2016)
54. Y. Song, K.W. Sung, Y. Han, Impact of packet arrivals on Wi-Fi and cellular system sharing unlicensed spectrum. IEEE Trans. Veh. Technol. **65**(12), 10204–10208 (2016)
55. Q. Chen, G. Yu, Z. Ding, Optimizing unlicensed spectrum sharing for LTE-U and WiFi network coexistence. IEEE J. Sel. Areas Commun. **34**(10), 2562–2574 (2016)
56. F. Tian, Y. Yu, X. Yuan, B. Lyu, G. Gui, Predicted decoupling for coexistence between WiFi and LTE in unlicensed band. IEEE Trans. Veh. Technol. **69**(4), 4130–4141 (2020)
57. M. Zhang, X. Zhang, Y. Chang, D. Yang, Dynamic uplink radio access selection of LTE licensed-assisted access to unlicensed spectrum: an optimization game. IEEE Commun. Lett. **20**(12), 2510–2513 (2016)
58. M.S. Ali, H. Tabassum, E. Hossain, Dynamic user clustering and power allocation for uplink and downlink non-orthogonal multiple access (NOMA) systems. IEEE Access **4**, 6325–6343 (2016)
59. Z. Wei, J. Guo, D.W.K. Ng, J. Yuan, Fairness comparison of uplink NOMA and OMA, in *Proceedings of the IEEE Vehicle Technical Conference (VTC)* (IEEE, New York, 2017), pp. 1–6
60. N. Zhang, J. Wang, G. Kang, Y. Liu, Uplink nonorthogonal multiple access in 5G systems. IEEE Commun. Lett. **20**(3), 458–461 (2016)
61. S.R. Islam, N. Avazov, O.A. Dobre, K.-S. Kwak, Power-domain non-orthogonal multiple access (NOMA) in 5G systems: potentials and challenges. IEEE Commun. Surv. Tutorials **19**(2), 721–742 (2017)
62. L. Dai, B. Wang, Y. Yuan, S. Han, I. Chih-Lin, Z. Wang, Non-orthogonal multiple access for 5G: solutions, challenges, opportunities, and future research trends. IEEE Commun. Mag. **53**(9), 74–81 (2015)
63. E. Hossain, Y. Al-Eryani, Large-scale NOMA: promises for massive machine-type communication (2019). arXiv preprint arXiv:1901.07106 .

64. R. Abbas, M. Shirvanimoghaddam, Y. Li, B. Vucetic, On the performance of massive grant-free NOMA, in *Proceedings of the IEEE International Symposi on Personal, Indoor and Mobile Radio Communication (PIMRC)* (IEEE, New York, 2017), pp. 1–6

65. Z. Wei, J. Yuan, D.W.K. Ng, M. Elkashlan, Z. Ding, A survey of downlink non-orthogonal multiple access for 5G wireless communication networks. arXiv preprint arXiv:1609.01856 (2016)

66. T. Lv, Y. Ma, J. Zeng, P.T. Mathiopoulos, Millimeter-wave NOMA transmission in cellular M2M communications for Internet of Things. IEEE Internet Things J. **5**(3), 1989–2000 (2018)

67. J. Zhu, J. Wang, Y. Huang, S. He, X. You, L. Yang, On optimal power allocation for downlink non-orthogonal multiple access systems. IEEE J. Sel. Areas Commun. **35**(12), 2744–2757 (2017)

68. F. Liu, P. Mähönen, M. Petrova, Proportional fairness-based power allocation and user set selection for downlink NOMA systems, in *Proceedings of the IEEE International Conference Communication (ICC), Kuala Lumpur, Malaysia* (2016)

69. Z. Ding, P. Fan, H.V. Poor, Impact of user pairing on 5G nonorthogonal multiple-access downlink transmissions. IEEE Trans. Veh. Technol. **65**(8), 6010–6023 (2016)

70. A. Celik, R.M. Radaydeh, F.S. Al-Qahtani, A.H.A. El-Malek, M.-S. Alouini, Resource allocation and cluster formation for imperfect NOMA in DL/UL decoupled hetnets, in *IEEE Globecom Workshops (GC Wkshps), Singapore* (2017)

71. M.A. Sedaghat, R.R. Müller, On user pairing in uplink NOMA. IEEE Trans. Wireless Commun. **17**(5), 3474–3486 (2018)

72. H. Haci, H. Zhu, J. Wang, A novel interference cancellation technique for non-orthogonal multiple access (NOMA), in *Proceedings of the IEEE Global Communication Conference (GLOBECOM), San Diego, CA, USA* (2015)

73. Z. Yang, W. Xu, H. Xu, J. Shi, M. Chen, Energy efficient non-orthogonal multiple access for machine-to-machine communications. IEEE Commun. Lett. **21**(4), 817–820 (2017)

74. Z. Yang, W. Xu, Y. Pan, C. Pan, M. Chen, Energy efficient resource allocation in machine-to-machine communications with multiple access and energy harvesting for IoT. IEEE Internet Things J. **5**(1), 229–245 (2017)

75. C.W. Sung, Y. Fu, A game-theoretic analysis of uplink power control for a non-orthogonal multiple access system with two interfering cells, in *Proceedings of the IEEE Vehicle Technical Conference (VTC), Nanjing, China* (2016)

76. X. Chen, A. Benjebbour, A. Li, A. Harada, Multi-user proportional fair scheduling for uplink non-orthogonal multiple access (NOMA), in *Proceedings of the IEEE Vehicle Technical Conference (VTC), Seoul, South Korea* (2014)

77. J. Choi, NOMA-based random access with multichannel Aloha. IEEE J. Sel. Areas Commun. **35**(12), 2736–2743 (2017)

78. E. Balevi, F.T. Al Rabee, R.D. Gitlin, ALOHA-NOMA for massive machine-to-machine IoT communication, in *Proceedings of the IEEE International Conference Communication (ICC), Kansas City, MO, USA* (2018)

79. M. Elkourdi, A. Mazin, E. Balevi, R.D. Gitlin, Enabling Aloha-NOMA for massive M2M communication in IoT networks (2018). arXiv preprint arXiv:1803.09513

80. H. Jiang, Q. Cui, Q. Gu, X. Qin, X. Zhang, X. Tao, Distributed layered grant-free non-orthogonal multiple access for massive MTC, in *Proceedings of the IEEE International Symposium on Personal, Indoor and Mobile Radio Commun. (PIMRC), Bologna, Italy* (2018)

81. J.-B. Seo, B.C. Jung, H. Jin, Performance analysis of NOMA random access. IEEE Commun. Lett. **22**(11), 2242–2245 (2018)

82. Z. Chen, Y. Liu, S. Khary, L.X. Cai, Y. Cheng, R. Zhang, Optimizing non-orthogonal multiple access in random access networks (2020). arXiv preprint arXiv:2002.06116

83. M. Qu, J. Liu, J.-B. Seo, H. Jin, *Distributed Fair Channel Access in NOMA Random Access Systems* (IEEE, New York, 2019), pp. 1–6

84. Y. Du, B. Dong, Z. Chen, X. Wang, Z. Liu, P. Gao, S. Li, Efficient multi-user detection for uplink grant-free NOMA: prior-information aided adaptive compressive sensing perspective. IEEE J. Sel. Areas Commun. **35**(12), 2812–2828 (2017)

85. B. Wang, L. Dai, Y. Zhang, T. Mir, J. Li, Dynamic compressive sensing-based multi-user detection for uplink grant-free NOMA. IEEE Commun. Lett. **20**(11), 2320–2323 (2016)
86. B. Wang, L. Dai, T. Mir, Z. Wang, Joint user activity and data detection based on structured compressive sensing for NOMA. IEEE Commun. Lett. **20**(7), 1473–1476 (2016)
87. F. Monsees, M. Woltering, C. Bockelmann, A. Dekorsy, Compressive sensing multi-user detection for multicarrier systems in sporadic machine type communication, in *Proceedings of the IEEE Vehicle Technical Conference (VTC), Glasgow, UK* (2015)
88. F. Monsees, C. Bockelmann, A. Dekorsy, Reliable activity detection for massive machine to machine communication via multiple measurement vector compressed sensing, in *IEEE Globecom Workshops (GC Wkshps)* (IEEE, New York, 2014), pp. 1057–1062
89. J. Choi, NOMA-based compressive random access using gaussian spreading. IEEE Trans. Commun. **67**(7), 5167–5177 (2019)
90. B. Wang, L. Dai, Y. Zhang, T. Mir, J. Li, Dynamic compressive sensing-based multi-user detection for uplink grant-free NOMA. IEEE Commun. Letters **20**(11), 2320–2323 (2016)
91. M. Shirvanimoghaddam, M. Condoluci, M. Dohler, S.J. Johnson, On the fundamental limits of random non-orthogonal multiple access in cellular massive IoT. IEEE J. Sel. Areas Commun. **35**(10), 2238–2252 (2017)
92. A.-S. Bana, E. De Carvalho, B. Soret, T. Abrão, J.C. Marinello, E.G. Larsson, P. Popovski, Massive MIMO for internet of things (IoT) connectivity. Phys. Commun. **37**, 100859 (2019)
93. F.A. De Figueiredo, F.A. Cardoso, I. Moerman, G. Fraidenraich, On the application of massive MIMO systems to machine type communications. IEEE Access **7**, 2589–2611 (2018)
94. N.H. Mahmood, H. Alves, O.A. López, M. Shehab, D.P.M. Osorio, M. Latva-aho, Six key enablers for machine type communication in 6G (2019). arXiv preprint arXiv:1903.05406
95. T.E. Bogale, L.B. Le, Massive MIMO and mmwave for 5G wireless hetnet: potential benefits and challenges. IEEE Veh. Technol. Mag. **11**(1), 64–75 (2016)
96. E. Björnson, E. De Carvalho, J.H. Sørensen, E.G. Larsson, P. Popovski, A random access protocol for pilot allocation in crowded massive MIMO systems. IEEE Trans. Wireless Commun. **16**(4), 2220–2234 (2017)
97. H. Han, X. Guo, Y. Li, A high throughput pilot allocation for M2M communication in crowded massive MIMO systems. IEEE Trans. Veh. Technol. **66**(10), 9572–9576 (2017)
98. J.C. Marinello, T. Abrão, Collision resolution protocol via soft decision retransmission criterion. IEEE Trans. Veh. Technol. **68**(4), 4094–4097 (2019)
99. J.C. Marinello, T. Abrão, R.D. Souza, E. de Carvalho, P. Popovski, Achieving fair random access performance in massive MIMO crowded machine-type networks. IEEE Wireless Commun. Lett. **9**(4), 503–507 (2019)
100. E. De Carvalho, E. Björnson, J.H. Sørensen, E.G. Larsson, P. Popovski, Random pilot and data access in massive MIMO for machine-type communications. IEEE Trans. Wireless Commun. **16**(12), 7703–7717 (2017)
101. J.H. Sørensen, E. De Carvalho, Č. Stefanovic, P. Popovski, Coded pilot random access for massive MIMO systems. IEEE Trans. Wireless Commun. **17**(12), 8035–8046 (2018)
102. Z. Chen, F. Sohrabi, W. Yu, Sparse activity detection for massive connectivity. IEEE Trans. Signal Process. **66**(7), 1890–1904 (2018)
103. L. Liu, W. Yu, Massive connectivity with massive MIMO—part I: device activity detection and channel estimation. IEEE Trans. Signal Process. **66**(11), 2933–2946 (2018)
104. K. Senel, E.G. Larsson, Grant-free massive MTC-enabled massive MIMO: a compressive sensing approach. IEEE Trans. Commun. **66**(12), 6164–6175 (2018)
105. S. Ali, A. Ferdowsi, W. Saad, N. Rajatheva, Sleeping multi-armed bandits for fast uplink grant allocation in machine type communications, in *IEEE Globecom Workshops (GC Wkshps)* (IEEE, New York, 2018), pp. 1–6

106. H.D. Trinh, L. Giupponi, P. Dini, Mobile traffic prediction from raw data using LSTM networks, in *Proceedings of the IEEE International Symposium on Personal, Indoor and Mobile Radio Communication (PIMRC)* (IEEE, New York, 2018), pp. 1827–1832

107. F. Xu, Y. Lin, J. Huang, D. Wu, H. Shi, J. Song, Y. Li, Big data driven mobile traffic understanding and forecasting: a time series approach. IEEE Trans. Serv. Comput. **9**(5), 796–805 (2016)

108. C. Qiu, Y. Zhang, Z. Feng, P. Zhang, S. Cui, Spatio-temporal wireless traffic prediction with recurrent neural network. IEEE Wireless Commun. Lett. **7**(4), 554–557 (2018)

109. J. Wang, J. Tang, Z. Xu, Y. Wang, G. Xue, X. Zhang, D. Yang, Spatiotemporal modeling and prediction in cellular networks: a big data enabled deep learning approach, in *Proceedings of the IEEE International Conference on Computer Communication (INFOCOM)* (IEEE, New York, 2017), pp. 1–9

110. C. Zhang, H. Zhang, D. Yuan, M. Zhang, Citywide cellular traffic prediction based on densely connected convolutional neural networks. IEEE Commun. Lett. **22**(8), 1656–1659 (2018)

111. L. Fang, X. Cheng, H. Wang, L. Yang, Mobile demand forecasting via deep graph-sequence spatiotemporal modeling in cellular networks. IEEE Internet Things J. **5**(4), 3091–3101 (2018)

112. S. Ali, N. Rajatheva, W. Saad, Fast uplink grant for machine type communications: challenges and opportunities. IEEE Commun. Mag. **57**(3), 97–103 (2019)

113. J. Brown, J.Y. Khan, A predictive resource allocation algorithm in the LTE uplink for event based M2M applications. IEEE Trans. Mobile Comput. **14**(12), 2433–2446 (2015)

114. S. Ali, W. Saad, N. Rajatheva, A directed information learning framework for event-driven M2M traffic prediction. IEEE Commun. Lett. **22**(11), 2378–2381 (2018)

115. Y. Wu, Y. Cui, W. Yu, C. Lu, W. Zhao, Modeling and forecasting of timescale network traffic dynamics in M2M communications, in *IEEE International Conference on Distributed Computing Systems (ICDCS)* (IEEE, New York, 2019), pp. 711–721

Chapter 3
MDP-Based Access Scheme for Virtualized M2M Networks

3.1 Introduction

Recently, the design of access schemes for MTC has received considerable attention [1–5]. There are several access schemes presented in the literature that assume that the network consists of only one application [6–9] while in the virtualized networks multiple applications or slices share the same infrastructure and the predetermined QoS, as requested by each slice, should be provided. Also, due to the small length of packets in these networks, schemes with large signaling overhead are not efficient [2, 10–12]. Furthermore, realistic arrival traffic models need to be considered for devices according to a variety of emerging services where it is not practical to assume the saturated scenario for devices.

Addressing such challenges, in this chapter, we aim to design an access scheme which takes into account the dynamic nature of arrival traffic of devices within slices [13]. We assume that the packet arrival state of devices evolves as a Markov model over frames and the AP is aware of the state transition probabilities. Under this assumption, a two-phase Markov decision process (MDP)-based access scheme is presented. In the first phase, a decision-making problem is formulated based on an MDP to develop an optimal polling mechanism. The design objective of this MDP is to maximize the network throughput subject to slice reservations by applying a pricing mechanism.

In the second phase, we present an access scheme to virtually realize the proposed decision-theoretic polling mechanism. To this end, first, the optimal access policy is studied by introducing the policy tree. Then, a heuristic algorithm is proposed for deterministic backoff generation based on the MDP formulation. The aim is to avoid the need for transmitting the policy tree or frequent polling packets. Through numerical results, we verify the performance of the developed access schemes in terms of packet delivery ratio, isolation, and throughput. In this chapter, we quantify isolation capability of an access scheme in a virtualized M2M network

© Springer Nature Switzerland AG 2020
T. Le-Ngoc, A. Dalili Shoaei, *Learning-Based Reconfigurable Multiple
Access Schemes for Virtualized MTC Networks*, Wireless Networks,
https://doi.org/10.1007/978-3-030-60382-3_3

in terms of the fairness among slices. More specifically, we measure such fairness in terms of equitable packet delivery ratios for different slices relative to their reservations. It is confirmed that the proposed approach noticeably improves the network throughput and provides better fairness among slices in comparison with TDMA, CSMA, and DEB.

The rest of this chapter is organized as follows. We first introduce the system model in Sect. 3.2. Then, Sect. 3.3 presents the MDP formulation and the optimal access policy. Subsequently, a heuristic access scheme is introduced in Sect. 3.4. Furthermore, Sect. 3.5 presents the numerical results. Finally, we conclude the chapter in Sect. 3.6.

3.2   System Model

We consider an M2M network with an AP which carries traffic belonging to S different slices. $\mathcal{D}_s$ denotes the set of devices subscribed to slice s, where $|\mathcal{D}_s| = N_s$ is the number of devices of slice s. The total number of all devices in all slices is N_d, where $N_d = \sum_{s=1}^{S} N_s$. Time is divided into fixed-length frames, each having a beacon followed by N_{ts} time-slots as shown in Fig. 3.1. In this setting, the smallest unit of time is called a time unit. Each time-slot has the duration of T_{ts}, which is equal to N_{tu} time units. $N_f = N_{ts} N_{tu}$ indicates the total number of time units in a frame. The beacon is broadcast by the AP to devices. It is assumed that each slice s can reserve r_s time-slots per frame.

The traffic generated at each device is described by an ON-OFF Markov chain, which is a common model for event-driven traffic patterns [14]. Each device d_s of slice s can be in two states at each frame t; $\phi_{d_s}^{pa}(t) = 0$ if it has no packet to transmit, and $\phi_{d_s}^{pa}(t) = 1$ otherwise. Device d_s transits from state 0 to state 1 with probability $a_{d_s}^0$ and stays in state 1 with probability $a_{d_s}^1$. During an ON period, it generates

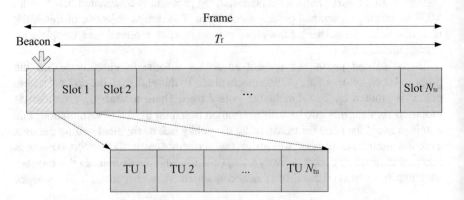

Fig. 3.1 Frame structure of the proposed MDP-based access scheme

constant bit rate (CBR) data, i.e., N_p packets per frame, each with a length of L_p. If
the device cannot access the channel before reception of the next beacon, the packet
will be dropped.

We assume that the wireless channel model includes path loss and small scale
fading. The received signal-to-noise ratio (SNR) of device d_s at the AP is equal to
$p_t|g_{d_s}|^2/\sigma_n^2$, where p_t is the transmission power, σ_n^2 is the noise power, and g_{d_s} is
the channel gain of the link from device d_s to the AP. More specifically, g_{d_s} is equal
to $cg'_{d_s}l_{d_s}^{-\zeta/2}$, where l_{d_s} is the distance between device d_s and the AP, ζ is the path
loss exponent, c is a constant dependent on the frequency and transmitter/receiver
antenna gain, and g'_{d_s} represents the small scale fading component. For simplicity,
without loss of generality, we normalize $cg'_{d_s} = 1$ in the following discussions.
If the received signal level falls below the receiver threshold, the receiver cannot
decode the signal successfully. The probability that the received SNR is less than
the receiver threshold ξ_R (i.e., outage probability) is given by

$$\psi_{d_s} = \mathbb{P}\left(\frac{p_t l_{d_s}^{-\zeta}}{\sigma_n^2} \leq \xi_R\right) = 1 - e^{-\frac{\sigma_n^2 l_{d_s}^\zeta \xi_R}{p_t}}. \tag{3.1}$$

Our main objective in this chapter is to develop an access scheme for a virtualized
M2M network that could maximize the network throughput, while considering
isolation among slices over each frame. To achieve these goals, we benefit both
from carrier sensing as in CSMA and deterministic backoff generation as in
DEB. In CSMA, since devices pick backoff values randomly, a collision may
occur if two devices pick the same backoff value. To avoid such collisions, in
[15], a deterministic backoff generation scheme is proposed in which non-equal
deterministic backoff values are assigned to devices by the AP. In this chapter, we
aim to use the idea of deterministic backoff generation to avoid collisions and ensure
isolation among slices. However, assuming that the AP is not aware of device states
at each frame and devices are not saturated, assigning backoff values in a round-
robin manner is not efficient. Thus, in the process of backoff value generation, we
take into account slice reservations and state transition probabilities of devices.

3.3 MDP-Based Access Scheme Design

Here, we propose a virtual polling-based access scheme for a virtualized M2M
network. First, we propose a decision-making problem to develop an optimal
polling mechanism. Second, we present access schemes to virtually realize the
proposed decision-theoretic polling mechanism with the aid of deterministic backoff
generation, while avoiding frequent polling overheads.

A. MDP Formulation To develop a polling mechanism, at each time unit in a frame,
the AP must determine which device to poll. Assume that the AP only knows the

statistical properties of arrival traffic (i.e., $a_{d_s}^0$ and $a_{d_s}^1$) and the instantaneous device states (i.e., $\phi_{d_s}^{pa}(t)$) at each frame are unknown. Since the AP has partial knowledge of the system, this decision-making problem can be formulated as an MDP with the following elements.[1]

(1) States Let $\phi_{d_s}(t, b)$ be the *transmission status* of device d_s in frame t and time unit b, where $b \in \{1, \ldots, N_f\}$. More specifically, $\phi_{d_s}(t, b) = 1$ indicates that device d_s has been already chosen to access the channel and transmission has happened successfully. $\phi_{d_s}(t, b) = -1$ represents that device d_s had an unsuccessful transmission due to outage. Furthermore, $\phi_{d_s}(t, b) = 0$ means that the device has been granted to access the channel but it has no packet for transmission. Finally, $\phi_{d_s}(t, b) = -2$ means that device d_s is not chosen for transmission. Thus, the system state can be represented by $\boldsymbol{\Phi}(t, b) = [\boldsymbol{\Phi}_s(t, b)]_{s=1}^S$, where $\boldsymbol{\Phi}_s(t, b) = [\phi_{d_s}(t, b)]_{d_s \in \mathcal{D}_s}$.

(2) Action The action corresponds to the device that is chosen to be granted channel access in frame t and time unit b, and is denoted by $d_a(t, b)$. Note that a ($a = 1, \ldots, S$) represents the index of the slice for the selected device for accessing the channel at t.

(3) Transition Function The transition function represented by $\mathbb{T}(\boldsymbol{\Phi}, d_a, \boldsymbol{\Phi}')$ gives the transition probability from state $\boldsymbol{\Phi}$ to $\boldsymbol{\Phi}'$ taking action d_a. At each frame, $\mathbb{T}(\boldsymbol{\Phi}, d_a, \boldsymbol{\Phi}')$ is zero except for

$$\mathbb{T}(\boldsymbol{\Phi}, d_a, \boldsymbol{\Phi}') = \begin{cases} \omega_{d_a}(1 - \psi_{d_s}), & \text{if } \phi_{d_a} = -2, \phi'_{d_a} = 1, \text{ and } \phi_i = \phi'_i, \forall i \neq d_a \\ \omega_{d_a}\psi_{d_s}, & \text{if } \phi_{d_a} = -2, \phi'_{d_a} = -1, \text{ and } \phi_i = \phi'_i, \forall i \neq d_a \quad (3.2) \\ 1 - \omega_{d_a}, & \text{if } \phi_{d_a} = -2, \phi'_{d_a} = 0, \text{ and } \phi_i = \phi'_i, \forall i \neq d_a, \end{cases}$$

where ω_{d_a} represents the belief value of device d_a at a given frame. In frame t, $\omega_{d_s}(t)$ is equal to the conditional probability (given the decision and transmission observation history) that device d_s has a packet to transmit and can be updated as

$$\omega_{d_s}(t) = \begin{cases} a_{d_s}^1, & \text{if } |\phi_{d_s}(t - 1, N_f)| = 1 \\ a_{d_s}^0, & \text{if } \phi_{d_s}(t - 1, N_f) = 0 \quad (3.3) \\ a_{d_s}^0(1 - \omega_{d_s}(t - 1)) + a_{d_s}^1\omega_{d_s}(t - 1), & \text{if } \phi_{d_s}(t - 1, N_f) = -2. \end{cases}$$

If the AP knows that device d_s had a packet in the last frame (i.e., $|\phi_{d_s}(t - 1, N_f)| = 1$), $\omega_{d_s}(t)$ is updated to $a_{d_s}^1$, which is the probability to stay in state 1. If the AP is aware that device d_s had no packet to transmit (i.e., $\phi_{d_s}(t - 1, N_f) = 0$), $\omega_{d_s}(t)$ is updated to $a_{d_s}^0$, which is the probability to transit from state 0 to 1. Otherwise, if the AP is not aware of device state (i.e.,

---

[1] An overview of MDP model is provided in Appendix.

$\phi_{d_s}(t - 1, N_f) = -2)$, $\omega_{d_s}(t)$ is updated to $a_{d_s}^0(1 - \omega_{d_s}(t - 1)) + a_{d_s}^1 \omega_{d_s}(t - 1)$, which represents the conditional probability of having a packet in frame t based on the law of total probability. A vector that consists of all belief values of different devices is represented by $\boldsymbol{\Omega}(t) = [\omega_{d_s}(t)]$, which is called the belief vector.

(4) Reward Function The reward gained by choosing device d_a to access the channel in a time unit is denoted by $\mathbb{R}(\boldsymbol{\Phi}, d_a, \boldsymbol{\Phi}')$, while the system state was at $\boldsymbol{\Phi}$. Considering the slice reservations, $\mathbb{R}(\boldsymbol{\Phi}, d_a, \boldsymbol{\Phi}')$ is defined as

$$\mathbb{R}(\boldsymbol{\Phi}, d_a, \boldsymbol{\Phi}') = (1 - \psi_{d_s})\mathbb{U}(\phi'_{d_a}) - \tag{3.4}$$

$$\gamma \left[\sum_{j \in \mathcal{D}_a} \mathbb{U}(\phi'_j) - r_a \right]^+ + \gamma \left[\sum_{j \in \mathcal{D}_a} \mathbb{U}(\phi_j) - r_a \right]^+,$$

where r_a is the time-slot reservation of slice a, which device d_a belongs to, γ is a positive scalar, $\mathbb{U}(x) = N_p L_p$ if $|x| = 1$, and $\mathbb{U}(x) = 0$ otherwise. Furthermore, $[x]^+ = \max\{x, 0\}$. In (3.4), the first component counts the packet transmission if the chosen device (i.e., device d_a) has a packet to transmit. The second and third components work as a pricing policy to avoid choosing a device whose slice has already been assigned a sufficient number of time-slots according to its reservation.

(5) Objective Function The design objective is to maximize the reward over one frame, which consists of N_f time units. Thus, in frame t and time unit b, a device is picked such that the expected total reward obtained over the remaining time of the frame is maximized. A policy $\pi_{t,b}(\boldsymbol{\Phi})$ is a mapping from state $\boldsymbol{\Phi}$ at time unit b in frame t to an action. The optimal policy is the one that achieves the highest expected reward. To obtain the optimal policy, we refer to the value function $\mathbb{V}_{t,b}(\boldsymbol{\Phi})$, which denotes the maximum expected remaining reward that can be accrued starting from $\boldsymbol{\Phi}$ and time unit b in frame t,

$$\mathbb{V}_{t,b}(\boldsymbol{\Phi}) = \max_{d_a} \sum_{\forall \boldsymbol{\Phi}' \neq \boldsymbol{\Phi}} \mathbb{T}(\boldsymbol{\Phi}, d_a, \boldsymbol{\Phi}') \left[\mathbb{R}(\boldsymbol{\Phi}, d_a, \boldsymbol{\Phi}') + \mathbb{V}_{t,b+\mathbb{H}(\boldsymbol{\Phi}, d_a, \boldsymbol{\Phi}')}(\boldsymbol{\Phi}') \right],$$

$$\tag{3.5}$$

where $\mathbb{H}(\boldsymbol{\Phi}, d_a, \boldsymbol{\Phi}') = L_p N_p$ if $\phi'_{d_a} = 1$ and $\mathbb{H}(\boldsymbol{\Phi}, d_a, \boldsymbol{\Phi}') = 1$ if $\phi'_{d_a} = 0$. Thus, the optimal policy can be obtained as

$$\pi^*_{t,b}(\boldsymbol{\Phi}) = \arg\max_{d_a} \mathbb{V}_{t,b}(\boldsymbol{\Phi}). \tag{3.6}$$

To solve (3.6), one approach is to use the value iteration technique whose computational complexity is exponentially growing with N_d [16]. To overcome this issue, the optimal action can be derived on the basis of the highest expected immediate reward instead of expected remaining reward. In other words, at each time unit, the effect of action on future decisions can be neglected. In the following, we consider this reduced-complexity approach by optimizing only the expected immediate reward.

B. Optimal Access Policy In the polling mechanism, the AP transmits the polling packet to the intended device. If the device has a packet for transmission, it transmits, otherwise, the AP sends a polling packet to another device. In the proposed access scheme, the aim is to eliminate the need for transmitting polling packets. To do that, in the beginning of each frame, the AP computes the order for devices to access the channel in that frame, and sends this information in the beacon. More specifically, first, it computes which device can access the channel in the first time unit. The chosen device may have a packet for transmission or not. For each of these possible cases, the AP chooses the next device to be granted channel access and the process continues.

The results of these series of computations can be represented by a tree, where each node indicates the transmission status of devices and its corresponding action, i.e., being in that state which device is granted to access the channel. From each node, at most two branches are originated; the left branch represents $\phi_{d_s}^{pa}(t) = 1$, i.e., the chosen device has a packet to transmit, and the right one represents $\phi_{d_s}^{pa}(t) = 0$ (see Fig. 3.2).

To access the channel with the help of information provided by the proposed policy tree, each device has to sense the channel from the first time unit and keep track of the channel state for each time unit. Consequently, at each stage, the device can learn the system transmission status. At each transmission state, the device that is chosen to transmit and has a packet for transmission, senses the channel for one time unit. If the channel is detected as idle, the device transmits its packet in the next time unit.

Here, we discuss the computational complexity of the algorithm to derive the policy tree. The computational complexity is calculated by counting the number of elementary operations performed by the algorithm, such as addition, subtraction, multiplication, division, comparison, etc. In this binary tree, assuming that root lies in stage 0, the first right branch terminates at stage of $\min(N_f - (N_p L_p + 1), N_d)$.

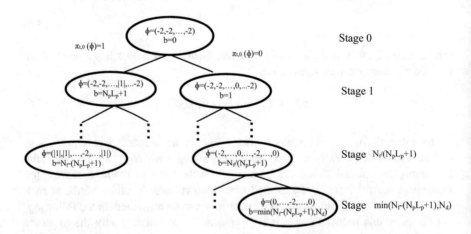

Fig. 3.2 Policy tree

Algorithm 1 Heuristic Algorithm for Backoff Generation

Output: $\boldsymbol{v}$ is the $1N_d$ vector indicating backoff value of all devices with element v_{d_s}
Initialization: $\rho = [\rho_s]_{s=1}^S = \mathbf{0}$, $\boldsymbol{v} = [-1]_1^{N_s}$,
$\kappa \leftarrow \text{sort}(\boldsymbol{\Omega}(t)(1 - \boldsymbol{\Phi})$, in descending order), $j = 0$
for $i = 1 : N_d$ **do**
    $d_s \leftarrow \kappa_i$, $\iota = (\omega_{d_s} N_p L_p + (1 - \omega_{d_s}))$
    If $\rho_s + \iota < r_s$ then $\rho_s = \rho_s + \iota$, $v_{d_s} = j$, and $j = j + 1$
end for
for each $d_s \in \mathcal{D}_s$, each $s = [1, 2, \ldots, S]$ **do**
    If $v(d_s) = -1$ then $v(d_s) = j$ and $j = j + 1$
end for

Thus, the total nodes of tree is $\mathcal{O}(2^{\min(N_d, N_f)})$, and each node requires $\mathcal{O}(N_d)$ operations to find the device which has the highest expected reward. Consequently, the total computational complexity of this algorithm is $\mathcal{O}(N_d 2^{\min(N_d, N_f)})$.

3.4 Heuristic Algorithm for Backoff Generation

For large values of N_d, the size of policy tree becomes large, which results in a high computational complexity and large beacons. Therefore, in this section, we present a simple but efficient algorithm to implement virtual polling through deterministic backoff generation. In the optimal policy, at each decision time, the device with the highest expected immediate reward is selected. According to (3.4), the reward is equal to the expected throughput if the number of time-slots allocated to each and every slice is smaller than its reservation. Thus, under this condition, the optimal solution would be to choose devices with the highest $\kappa_{d_s} = (1 - \psi_{d_s})\omega_{d_s}$. Based on this intuition, to reduce complexity, we propose a heuristic algorithm as follows.

At first, devices are sorted according to their κ_{d_s} in descending order. Then, at each step, we assign backoff values to devices which are started 0 and incremented by 1. The device with the highest κ_{d_s} will be selected at each step as long as the total time allocated to devices belonging to the slice s is smaller than r_s. However, since the moment that a selected slice is assigned a sufficient number of time-slots (equal to its reservation), the devices belonging to that slice will not be considered for backoff value allocation until all slices have met their reservations. This is because the slice which its allocated time-slots exceed its reservations will be penalized according to (3.4). Thus, next, the device from other slices with the highest κ_{d_s} among the remaining devices will be chosen. When the number of allocated time-slots for all slices exceeds their reservations, the backoff value assignment for the remaining devices will continue according to their κ_{d_s} in descending order.

In order to access the channel in this scheme, at each time unit, if the backoff value is equal to 0, the device would sense the channel for one time unit and transmit in the next time unit. In case of busy channel, the packet would be dropped. Otherwise, it would decrease its backoff value by 1. If the channel is sensed busy, the device would go to the sleep mode for L_p time units. Therefore, the device has to sense all time units smaller than its backoff value except for those that it is in the sleep mode. It should be noted that the complexity of this algorithm is dependent on the sorting algorithm which is $\mathcal{O}(N_d \log N_d)$. In the proposed deterministic scheme, each device is assigned a unique backoff value. The range of these values is from 1 to N_d. Therefore, backoff values can be represented by $\log N_d$ bits. Thus, the total bits required to pass backoff values to devices is $N_d \log N_d$.

A drawback of this algorithm is that it requires devices being able to synchronously count time units. However, a device's backoff counter may possibly become out of synchronization due to the hardware clock imperfections inside the wireless network interface cards, which is commonly referred to as clock drift. If a device loses synchronization with others, it would drift to a new time unit (other than the scheduled one). This time unit can be possibly busy and collision can happen. Nowadays, this event, i.e., the clock drift happening is less of a concern as it rarely occurs for commercial products [15].

3.5 Numerical Results

Here, we present numerical results to evaluate the performance of the proposed MDP-based access schemes in comparison with CSMA, DEB, and TDMA schemes in terms of *packet delivery ratio*, *isolation*, and *throughput*. For TDMA, it should be noted that time-slots are first distributed among slices based on their corresponding reservations. Within each slice s at each frame, r_s devices with the highest κ_{d_s} would be allocated to the time-slots. All algorithms are implemented in Matlab and the results are based on the average of 100 repeated simulations over random distributions of devices, each of which is 10,000 frames long. It should be noted that based on simulation results the standard deviation of the sample mean is less than 1% when sample size is 10,000 frames. We consider a virtualized M2M network serving 4 slices in a circular area with a radius of 5 m. Devices are randomly distributed (from a uniform distribution) in this area. Each slice has a reservation of 4 time-slots (i.e., $r_s = 4$). Furthermore, $N_{ts} = 16$, $L_p = T_{ts} = 12$ time units, $N_p = 1$, and $\gamma = 0.8$.[2] The channel parameters are set as path loss exponent $\zeta = 3$, receiver threshold $\xi_R = 0$ dB, and $\frac{P_t}{\sigma_n^2} = 20$ dB.

To study the impact of device distribution on the isolation and throughput achieved by the access schemes, we consider an unsaturated network with two examples of balanced and unbalanced device distributions. For a *balanced scenario*, all slices have the same number of devices (i.e., N_s) and arrival traffic statistics (i.e., a^0 and a^1). Let $N_s^a = N_s \lambda_1$ be the average number of active devices for slice s, where $\lambda_1 = \frac{a^0}{1-a^1+a^0}$ is the steady state probability of a device

---

[2]Note that larger values of γ lead to larger isolation indexes and lower network throughput.

being at state 1. We set $r_s = N_s^a N_p L_p$. This assumption represents a network, where its *average* load is equal to its capacity. But, it should be considered that the *instantaneous* number of active devices at each frame could be larger or smaller than the reservation. Accordingly, we consider 4 different values of $(a^0, a^1) \in \{(0.1, 0.9), (0.2, 0.6), (0.3, 0.1), (0.1, 0.6)\}$ for devices. For each value of (a^0, a^1), we set $N_s = r_s/(\lambda_1 N_p L_p)$, which results in $\{8, 12, 16, 20\}$. For an *unbalanced scenario*, one slice has a larger number of devices than other slices. This scenario represents an overloaded network. For this scenario, we consider $(a^0, a^1) = (0.1, 0.9)$ for all devices. Then, we set $N_s = 8$ for $s \in \{1, 2, 3\}$ which ensures $r_s = N_s^a N_p L_p$, while N_4 can accept larger values. Thus, the average number of active devices in slice 4 is larger than other slices and its own reservation.

(1) Packet Delivery Ratio Packet delivery ratio (PDR) is defined as the ratio of number of packets that are successfully transmitted to total number of generated packets. Figure 3.3a shows the PDR versus the number of devices per slice for the balanced scenario. As can be observed, the optimal and heuristic approaches achieve the same PDR. The reason is that here to obtain the optimal policy 3.6, only expected immediate reward is considered while the expected remaining reward over the frame is neglected. Comparing to CSMA, DEB, and TDMA, the proposed approaches improve the PDR performance. Nevertheless, it is shown that their performance is decreasing as N_s increases, although the average number of active devices is fixed to 4. The reason is that the size of backoff values assigned to the devices are dependent on the number of devices. With a larger N_s, larger backoff values would be assigned to devices which obviously causes more wasted time in waiting. On the other hand, the performance of CSMA is independent of N_s considering that N_s^a is fixed. This is because of the random backoff generation procedure in CSMA, which is affected by the number of active devices, not by the total number of devices.

Unlike CSMA, which is only affected by N_s^a, the TDMA performance is more sensitive to the arrival traffic statistics. In particular, its PDR is generally decreased with λ_1 since the probability that an assigned time-slot remains idle is decreased as shown in Fig. 3.3a. However, it is shown that the TDMA performance increases for $N_s = 20$ (corresponding to $(a^0, a^1) = (0.1, 0.6)$) compared to $N_s = 16$ (corresponding to $(a^0, a^1) = (0.3, 0.1)$). This can be explained by the fact that TDMA performance depends on both λ_1 and $|a^0 - a^1|$. The reason is that the larger value of $|a^0 - a^1|$ provides higher probability to predict the device state given observation of device state. Therefore, the probability that an assigned time-slot remains idle would be decreased.

Figure 3.3b illustrates PDR versus N_4, for the unbalanced scenario. As expected, for all access schemes, PDR decreases by increasing N_4. This is because the number of generated packets of slice 4 is increasing with N_4. Moreover, compared to the balanced scenario, it is shown that the PDR improvement provided by the proposed access schemes relative to DEB is increased. This fact shows the effectiveness of using the belief vector in backoff value assignment instead of round-robin assignment in DEB.

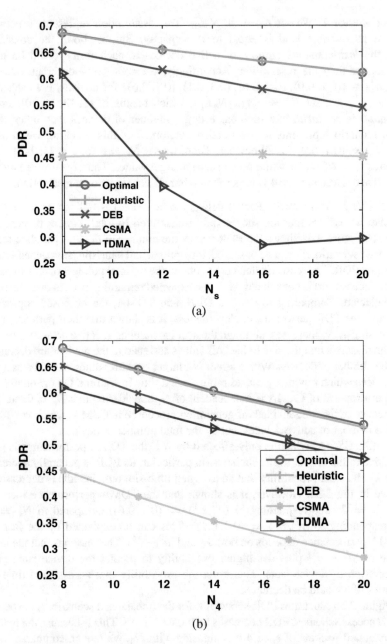

Fig. 3.3 Total packet delivery ratio. (**a**) PDR versus N_s, balanced scenario. (**b**) PDR versus N_4, unbalanced scenario

(2) Isolation Index We introduce an index to measure isolation among slices based on Jain's fairness index [17]. To this end, first, we define ρ_s, the PDR of slice s considering its reservation, as

$$\rho_s(t) = \min\left(\boldsymbol{\Gamma}_s(t)\boldsymbol{\Phi}_s^{\mathrm{pa}}(t)^\mathsf{T} / \min(\boldsymbol{\Phi}_s^{\mathrm{pa}}(t)\mathbf{1}^\mathsf{T}, r_s), 1\right), \tag{3.7}$$

where $\boldsymbol{\Phi}_s^{\mathrm{pa}}(t) = [\phi_{d_s}^{\mathrm{pa}}(t)]_{d_s \in \mathcal{D}_s}$ represents the device states and $\boldsymbol{\Gamma}_s(t)$ denotes the allocation strategy for slice s in frame t. Thus, $\boldsymbol{\Gamma}_s(t)\boldsymbol{\Phi}_s^{\mathrm{pa}}(t)^\mathsf{T}$ represents the number of transmitted packets for slice s in frame t. It should be noted that we limit the maximum number of generated packets of slice s (i.e. $\boldsymbol{\Phi}_s^{\mathrm{pa}}\mathbf{1}^\mathsf{T}$) that potentially can be served, to its reservation r_s. This is the reason $\min(\boldsymbol{\Phi}_s^{\mathrm{pa}}\mathbf{1}^\mathsf{T}, r_s)$ is used in the denominator in (3.7). In an ideal case, an access scheme would assign time to different slices such that $\rho_s(t) = 1$, regardless of number of active devices. Thus, we measure the isolation level provided by an access scheme based on the Jain's fairness index, which can represent the variability among $\rho_s(t)$, $\forall s$. We define the isolation index $I(t)$ in frame t as $I(t) = (\sum_{s=1}^{S} \rho_s)^2 / (S \sum_{s=1}^{S} \rho_s^2)$, where larger values of I indicates better isolation among different slices. Maximum value of I is 1, which indicates that all slices fairly have a time share of a frame relative to their reservations.

In Fig. 3.4a, I is demonstrated versus the number of devices per slice for the balanced scenario. Note that I is measured for each frame and then the average over 10,000 frames is plotted. It can be seen that the isolation index of the optimal policy is close to one. Except for CSMA, the isolation index decreases for all access schemes as N_s increases. Similar to PDR, the isolation index of TDMA is dependent on $|a^0 - a^1|$. Thus, it is increasing from $N_s = 16$ (corresponding to $(a^0, a^1) = (0.3, 0.1)$) to $N_s = 20$ (corresponding to $(a^0, a^1) = (0.1, 0.6)$).

Figure 3.4b shows the isolation index versus the number of devices belonging to slice 4 for the unbalanced scenario. As can be observed, for the proposed approaches, the isolation index is not affected by increasing N_4, whereas the isolation index of CSMA and DEB quickly drops when N_4 increases. This confirms the advantages of the proposed access schemes to manage isolation comparing to CSMA and DEB, that are incapable of handling a slice load imbalance situation in the network.

(3) Throughput Here, throughput is defined as the number of successfully transmitted packets per frame. For the unbalanced scenario, Fig. 3.5 illustrates the total throughput of all slices considering different access schemes. As can be observed, the proposed optimal and heuristic approaches outperform TDMA, DEB and CSMA. Furthermore, except for CSMA, it is shown that the throughput is increasing with N_4. The reason is that when the number of active devices increases, the chance of having a small number of packets compared to the capacity of the frame decreases. On the other hand, a larger number of devices leads to larger backoff values, which may cause more time wasted for backoff. However, since optimal, heuristic and TDMA schemes use the belief vector and $|a^0 - a^1| = 0.8$ is large, active devices are more likely assigned lower backoff values. However, the CSMA throughput decreases due to the higher number of collisions.

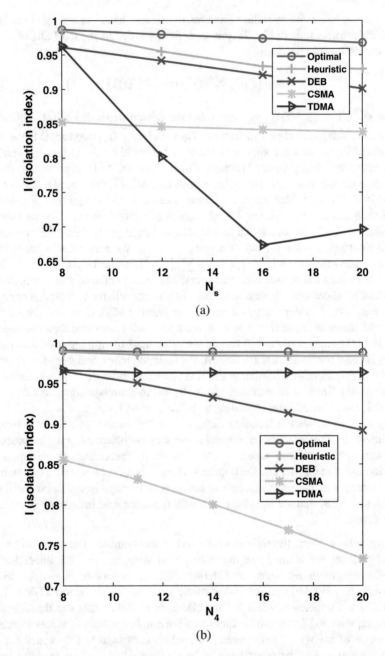

Fig. 3.4 Average isolation index. (**a**) Isolation versus N_s, balanced scenario. (**b**) Isolation versus N_4, unbalanced scenario

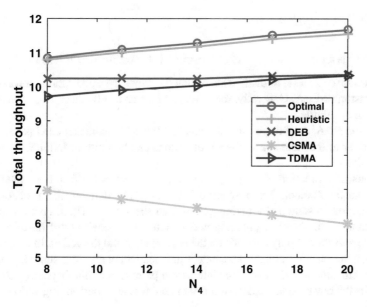

Fig. 3.5 Total throughput for $a^0 = 0.1$, $a^1 = 0.9$, and $N_1 = N_2 = N_3 = 8$

3.6   Concluding Remarks

In this chapter, we have proposed an MDP-based access scheme with deterministic backoffs for a virtualized M2M network to improve both network utilization and slice isolation. This approach works as a virtual polling-based access scheme, but without polling packet overheads. In this approach, at each frame, the AP assigns unique backoff values to devices based on their traffic statistics and slice reservations. Such deterministic backoff value assignment avoids collisions among devices. Numerical results confirm its efficiency in comparison with TDMA, DEB, and CSMA in unsaturated networks. The performance is measured in terms of PDR, isolation index, and throughput. It is shown that this MDP-based access scheme can keep isolation among different slices regardless of their numbers of devices or arrival traffic statistics.

Appendix: An overview of MDP model

In this section, we first provide the model description of MDP and next we discuss how this problem is formulated.

Model Description

An MDP is a tuple that consists of 5 components described as follows [18–22].

- State space S: a set of distinct states that describes the information relevant to the decision process. Generally, the number of states can be discrete or continuous, finite or infinite.
- Action set A: a set of possible actions that the decision maker can take at state s. Similar to the state space, the set of actions can be finite or infinite, continuous or discrete.
- Transition function T: captures the fact that actions affect the system in a stochastic manner. More specifically, $T(s', a, s)$ denotes the probability of transition to state s' by taking action a in the state s. The transition function represents the Markov property which states the probability of transition from state s to state s' only depends on the current state and the action. In other words, given the current state, the next state is independent of previous states and actions.
- Reward function R: maps the state-action pairs to real numbers. $R(s, a)$ represents the reward that the decision maker achieves after performing action a in the state s.

Time Horizon and Discount Factor

The objective of solving an MDP is to maximize the expected cumulative reward obtained by the decision maker over some time steps. The occurrence of an event, such as the performance of a stochastic action, constitutes one step. The horizon denotes how many time steps the decision maker will act which can be finite or infinite. Finite horizon denotes that the decision maker acts only finitely many time steps, while infinite horizon denotes that the decision maker acts forever in the system. The future reward can be discounted over time or averaged. In the discounted horizon formulation, a discount factor is used to weight the rewards obtained over time steps. The rewards will be be scaled down by steps from any starting point, and the overall expected reward is the sum of these discounted rewards over the horizon. In the average horizon formulation, the expected reward is computed as the average of total rewards obtained by the decision maker over horizon.

Observations, Policies and Value Function

In MDPs, the decision maker needs to choose actions based on observing the state that the decision maker is. Consequently, a policy is defined as plan that specifies which action the decision maker should taken at the current state based on its observation. Policies can be deterministic or non-deterministic, stationary or non-stationary. If the choices of actions are independent of the steps, the policy is stationary, otherwise, it is non-stationary. Furthermore, the policy is deterministic, if for any current state, only one action can be chosen, otherwise, it is non-deterministic.

Every policy is associated with a value function denoted by $V : S \longrightarrow \mathcal{R}$. The value function, for a fixed policy π, gives the expected cumulative reward starting from state s, which can be formally expressed as

$$V^{\pi}(s) = \mathcal{E}_{\pi}\{\sum_{t=0}^{n} \gamma_t R(s)|s\} \quad (3.8)$$

The value function can be expanded as

$$V_n^{\pi}(s) = R(s, \pi(s)) + \gamma \sum_{s \in S} T(s, \pi(s), t) V_{n-1}^{\pi}(t) \quad (3.9)$$

There are several ways to compute the optimal policy π^* such that $V^{\pi^*}(s) \geq V^{\pi}(s)$ for any $s \in S$ and any policy π. Since in all these techniques, a value function is used, we refer to them as value-based solvers [22].

In order to solve MDPs, value-based solvers deploy an iterative approach, where each iteration consists of two phases: (1) policy evaluation and (2) policy improvement, shown in Fig. 3.6. In the policy evaluation phase, the solver obtains the value function for some or all states given the fixed policy. In the policy improvement step, the algorithm improves the previous policy based on values calculated in the policy evaluation step. The process to evaluate and improve the policy continues until either no improvement is made in policy, a time limit has been

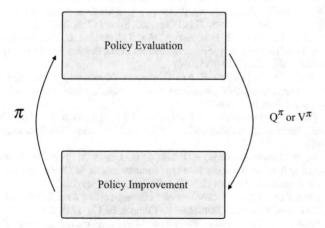

Fig. 3.6 Policy evaluation/improvement iteration

reached, or the change to the value function is below a certain threshold. Figure 3.6 does not show exactly how these phases are completed or when an algorithm switches between them. For example, in some algorithms (like policy iteration) the evaluation phase takes more time in others (like value iteration) the transition between phases happens quickly.

References

1. A. Laya, L. Alonso, J. Alonso-Zarate, Is the random access channel of LTE and LTE-A suitable for M2M communications? A survey of alternatives. IEEE Commun. Surv. Tutorials **16**(1), 4–16 (2014)
2. A. Rajandekar, B. Sikdar, A survey of MAC layer issues and protocols for machine-to-machine communications. IEEE Internet Things J. **2**(2), 175–186 (2015)
3. E. Soltanmohammadi, K. Ghavami, M. Naraghi-Pour, A survey of traffic issues in machine-to-machine communications over LTE. IEEE Internet Things J. **3**(6), 865–884 (2016)
4. F. Ghavimi, H.-H. Chen, M2M communications in 3GPP LTE/LTE-A networks: architectures, service requirements, challenges, and applications. IEEE Commun. Surv. Tutorials **17**(2), 525–549 (2014)
5. A. Biral, M. Centenaro, A. Zanella, L. Vangelista, M. Zorzi, The challenges of M2M massive access in wireless cellular networks. Digital Commun. Netw. **1**(1), 1–19 (2015)
6. H. He, Q. Du, H. Song, W. Li, Y. Wang, P. Ren, Traffic-aware ACB scheme for massive access in machine-to-machine networks, in *Proceeding of the IEEE International Conference Communication (ICC), London, UK* (2015)
7. W. Zhan, L. Dai, Throughput optimization for massive random access of M2M communications in LTE networks, in *Proceedings of the IEEE International Conference Communication (ICC), Paris, France* (2017)
8. S. Duan, V. Shah-Mansouri, Z. Wang, V.W. Wong, D-ACB: Adaptive congestion control algorithm for bursty M2M traffic in LTE networks. IEEE Trans. Veh. Technol. **65**(12), 9847–9861 (2016)
9. Y. Yang, S. Roy, Grouping-based MAC protocols for EV charging data transmission in smart metering network. IEEE J. Sel. Areas Commun. **32**(7), 1328–1343 (2014)
10. S. Chen, R. Ma, H.-H. Chen, H. Zhang, W. Meng, J. Liu, Machine-to-machine communications in ultra-dense networks—A survey. IEEE Commun. Surveys Tuts. **19**(3), 1478–1503 (2017)
11. C. Bockelmann, N. Pratas, H. Nikopour, K. Au, T. Svensson, C. Stefanovic, P. Popovski, A. Dekorsy, Massive machine-type communications in 5G: Physical and MAC-layer solutions. IEEE Commun. Mag. **54**(9), 59–65 (2016)
12. A. Rico-Alvarino, M. Vajapeyam, H. Xu, X. Wang, Y. Blankenship, J. Bergman, T. Tirronen, E. Yavuz, An overview of 3GPP enhancements on machine to machine communications. IEEE Commun. Mag. **54**(6), 14–21 (2016)
13. A. Dalili Shoaei, M. Derakhshani, S. Parsaeifard, T. Le-Ngoc, MDP-based MAC design with deterministic backoffs in virtualized 802.11 WLANs. IEEE Trans. Veh. Technol. **65**(9), 7754–7759 (2016)
14. N. Nikaein, M. Laner, K. Zhou, P. Svoboda, D. Drajic, M. Popovic, S. Krco, Simple traffic modeling framework for machine type communication, in *International Symposium on Wireless Communication Systems (ISWCS), Ilmenau, Germany* (2013)
15. Y. He, X. Ma, Deterministic backoff: toward efficient polling for IEEE 802.11 e HCCA in wireless home networks. IEEE Trans. Mobile Comput. **10**(12), 1726–1740 (2011)
16. M.L. Puterman, *Markov Decision Processes: Discrete Stochastic Dynamic Programming* (Wiley, New York, 2009)

17. R. Jain, D.-M. Chiu, W.R. Hawe, *A Quantitative Measure of Fairness and Discrimination for Resource Allocation in Shared Computer System* (Eastern Research Laboratory, Digital Equipment Corporation Hudson, Maynard, 1984)
18. A.G. Barto, S.J. Bradtke, S.P. Singh, Learning to act using real-time dynamic programming. Artif. Intell. **72**(1–2), 81–138 (1995)
19. C. Boutilier, R. Dearden, M. Goldszmidt et al., Exploiting structure in policy construction, in *IJCAI*, vol. 14 (1995), pp. 1104–1113
20. C. Boutilier, R. Dearden, Using abstractions for decision-theoretic planning with time constraints, in *AAAI* (1994), pp. 1016–1022
21. T.L. Dean, L.P. Kaelbling, J. Kirman, A.E. Nicholson, Planning with deadlines in stochastic domains, in *AAAI*, vol. 93 (1993), pp. 574–579
22. N. Roy, J. How, *A Tutorial on Linear Function Approximators for Dynamic Programming and Reinforcement Learning* (2013)

Chapter 4
Reconfigurable and Traffic-Aware Access Schemes for Virtualized M2M Networks

4.1 Introduction

In Chap. 3, we have proposed an access scheme using a carrier sense-based approach. In this scheme, instead of each device generating a random backoff to access the channel, backoff values are assigned by the AP. This approach achieves high throughput efficiency for scenarios consisting of devices with high traffic arrival probabilities, however its performance decreases with increasing the low-traffic devices. Furthermore, for scenarios the size of data packets is small [1–4], the time left idle for the carrier sensing purposes becomes large compared to the time used for data transmission, leading to low resource utilization. Moreover, it may lead to starvation for devices with a low probability of packet transmission.

In fact, in environments consisting of heterogeneous devices, higher efficiency and fairness can be achieved by adaptively switching between scheduling-based and random access schemes, which respectively show better efficiency for high and low-traffic devices [5, 6]. To deploy such a reconfigurable access scheme, the traffic statistics should be provided to the AP. In particular, the traffic statistics of the devices are used to properly select and configure an access scheme to maximize the spectral efficiency.

In this chapter, first, aiming to improve the network efficiency, we design a reconfigurable access scheme with optimal demand-free assignment-based (DFA) and random access-based (RA) partition according to the device packet arrival statistics [7, 8]. To this end, we formulate an optimization problem which is inherently non-convex and suffers from high computational complexity. To tackle this issue, we first show that the problem belongs to the class of complementary geometric programming (CGP). Then, we propose an efficient and tractable iterative approach to solve it. At each iteration, via applying transformation techniques and arithmetic geometric mean approximation (AGMA), we transform the CGP-based

© Springer Nature Switzerland AG 2020
T. Le-Ngoc, A. Dalili Shoaei, *Learning-Based Reconfigurable Multiple Access Schemes for Virtualized MTC Networks*, Wireless Networks, https://doi.org/10.1007/978-3-030-60382-3_4

formulation into a geometric programming (GP) form, which can be solved by softwares such as CVX efficiently [9].

Second, we propose a scalable solution for a large number of devices in M2M networks with considerably less computational complexity as compared to the proposed CGP-based scheduling. In particular, to overcome the computational burden caused by a large number of devices, the optimization problem is transformed using approximations for RA throughput and airtime. Subsequently, we propose an efficient iterative algorithm to solve the approximated optimization problem, where each iteration is decomposed into two sub-problems: one belongs to the linear-programming category, and the other is of the difference of convex (DC)-programming type.

The rest of this chapter is organized as follows. We first introduce the system model under consideration in Sect. 4.2. Section 4.3 presents the problem formulation. Subsequently, an iterative CGP-based algorithm along with a scalable reconfigurable access scheme are proposed in Sect. 4.4. Section 4.5 provides simulation results. Finally, Sect. 4.6 draws the conclusion.

4.2   System Model

4.2.1   Network Model and Frame Structure

We consider a single AP serving N_d devices, where all communications are done through the AP. Each device is exclusively subscribed to one slice with a specific airtime reservation. There are $S = \{1, \cdots, S\}$ different slices and $\mathcal{D}_s$ denotes the set of subscribed devices to slice s, where $|\mathcal{D}_s| = N_s$ is the number of devices at slice s.

Time is divided into fixed-length frames indexed by t. As shown in Fig. 4.1, each frame begins with a beacon issued by the AP followed by the DFA segment with a duration of less than $T_{\max}$ for scheduled devices, and the RA segment of the length $T_{ra}(t)$. In the DFA segment, the assignment is called demand assignment if the scheduled device has explicitly announced in the previous frames that its

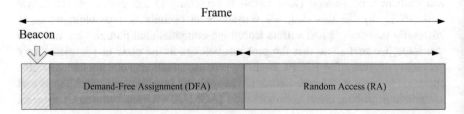

Fig. 4.1 Frame structure of the proposed reconfigurable access scheme for a virtualized M2M network

queue is not empty, otherwise the assignment is called free assignment. During the RA segment, the p-persistent CSMA scheme is used, where each time-slot with a duration of T_{ts} is divided into time units.

Regarding the required airtime per slice, it is assumed that each slice s can reserve time for r_s time-slots per frame. For delay-sensitive devices, an exclusive time-share in the DFA segment could be further reserved to meet their delay requirements.

4.2.2 Traffic and Channel Models

We consider Bernoulli process to model the packet arrivals of each device [10] since the traffic generated by devices is mainly sporadic with a negligible probability that more than one packet arrive in one frame. More specifically, we assume that a new packet is generated at device d_s with a probability of a_{d_s} at each frame and it is added to the queue of the device if the current length of the queue is smaller than Q_{max}. Otherwise, the packet is discarded. Furthermore, we first assume that the AP is aware of the vector of packet arrival probabilities of devices, represented by $A = [a_{d_s}]_{\forall d_s}$, and it keeps a vector denoted by $V(t) = [v_{d_s}(t)]_{\forall d_s}$, where v_{d_s} denotes the last time that the AP has received a packet from the device d_s. Moreover, each time the device sends a packet, it piggybacks an extra bit (denoted by $q_{d_s}(t)$) telling whether its queue is empty ($q_{d_s}(t) = 0$) or non-empty ($q_{d_s}(t) = 1$, i.e., it has packets backlogged in the queue to transmit). Therefore, at each frame t, the AP updates $\theta_{d_s}(t)$ that indicates the probability of the device d_s having a non-empty queue at t as

$$\theta_{d_s}(t) = \begin{cases} 1 - (1 - a_{d_s})^{t - v_{d_s}(t)}, & \text{if } q_{d_s}(v_{d_s}(t)) = 0 \\ 1 & \text{if } q_{d_s}(v_{d_s}(t)) = 1. \end{cases} \tag{4.1}$$

Regarding the wireless channel model, we consider path loss and small scale fading as described in Chap. 3, Sect. 3.2. In this model, whenever a device d_s belonging to slice s transmits a packet, with probability ψ_{d_s}, its received SNR at the AP falls below the required threshold.

4.2.3 Device Operation

Before the frame t starts, the AP decides on time-slot allocation for the DFA segment and notifies the schedule to the devices via the beacon. Devices with no allocated time-slot will attempt to transmit in the RA segment if they have a packet, using the p-persistent CSMA protocol as follows. In the RA segment, device d_s with a non-empty queue performs channel sensing. If the channel is sensed to be idle, device d_s transmits the packet with probability p_{d_s} at the beginning of the next time-slot or defers with probability $(1 - p_{d_s})$. If a device is unsuccessful in transmission of a packet either in DFA or RA segment, it is not allowed to retransmit the packet in that current frame. The reason is that, the retransmissions may affect the airtime of other slices.

4.2.4   An Analytical Model for p-persistent CSMA

Here, we model the throughput of p-persistent CSMA protocol in an unsaturated mode. Let P_{idle} be the probability that the channel is idle in a time unit. This probability is calculated as

$$P_{\text{idle}} = \prod_{s \in \mathcal{S}} \prod_{d_s \in \mathcal{D}_s} (1 - P_{\text{idle}}^{d_s}) = \prod_{s \in \mathcal{S}} \prod_{d_s \in \mathcal{D}_s} (1 - \theta_{d_s} p_{d_s}), \qquad (4.2)$$

where $P_{\text{idle}}^{d_s}$ represents the probability that device d_s is idle in a time unit and $\theta_{d_s} p_{d_s}$ denotes the transmission probability of device d_s. The key approximation is that we assume that the number of active devices is constant over a frame, however in the proposed scheme, the device transmits at most one packet during the RA segment.

A transmitted packet will be received successfully, if exactly one device transmits on the channel, and outage does not happen. For device d_s, the probability of *successful* transmission denoted by $P_{\text{succ}}^{d_s}$ is

$$P_{\text{succ}}^{d_s} = \theta_{d_s} p_{d_s} (1 - \psi_{d_s}) \prod_{s \in \mathcal{S}} \prod_{d_s' \in \mathcal{D}_s, d_s' \neq d_s} (1 - \theta_{d_s'} p_{d_s'}). \qquad (4.3)$$

As introduced in [11], the normalized throughput of device d_s (denoted by ρ_{d_s}) is defined as the fraction of time that the channel is used for its successful transmission,

$$\rho_{d_s} = \frac{P_{\text{succ}}^{d_s} T_{\text{succ}}}{P_{\text{idle}} \rho + (1 - P_{\text{idle}}) T_{\text{succ}}}, \qquad (4.4)$$

where ρ is the duration of a time unit and T_{succ} is the duration of a successful transmission, which includes the data transmission for a fixed time, inter-frame spaces, and signaling overheads. Since signaling and inter-frame spaces are relatively small (in the order of μs) compared with the data transmission length (in the order of ms), we approximately assume that both collided and successful transmissions are of the same size (i.e., T_{succ}). Furthermore, we assume that each transmission occupies 1 time-slot, i.e., $T_{\text{succ}} = T_{\text{ts}}$. Consequently, the denominator in (4.4) represents the expected length of a general time-slot.

By introducing a new variable, y_{d_s} as

$$y_{d_s} = \frac{\theta_{d_s} p_{d_s}}{1 - \theta_{d_s} p_{d_s}}, \tag{4.5}$$

we can simplify (4.4). To this end, first, we rewrite P_{idle} and $P_{\text{succ}}^{d_s}$ in terms of y_{d_s} as

$$P_{\text{idle}} = \frac{1}{\prod_{s \in S} \prod_{d_s \in \mathcal{D}_s} (1 + y_{d_s})}, \tag{4.6}$$

$$P_{\text{succ}}^{d_s} = \frac{y_{d_s} (1 - \psi_{d_s})}{\prod_{s \in S} \prod_{d_s \in \mathcal{D}_s} (1 + y_{d_s})} = y_{d_s} (1 - \psi_{d_s}) P_{\text{idle}}. \tag{4.7}$$

Then, we obtain ρ_{d_s} in terms of y_{d_s} as

$$\rho_{d_s} = \frac{y_{d_s}}{\prod_{s \in S} \prod_{d_s \in \mathcal{D}_s} (1 + y_{d_s}) - t'}, \tag{4.8}$$

where $t' = \frac{T_{\text{ts}} - \rho}{T_{\text{ts}}}$. Furthermore, in the context of virtualized wireless networks, *total access airtime* is considered as another performance metric to measure and preserve isolation. For device d_s, the total access airtime during the RA segment is defined as

$$\tau_{d_s} = \frac{(1 - P_{\text{idle}}^{d_s}) T_{\text{ts}}}{P_{\text{idle}} \rho + (1 - P_{\text{idle}}) T_{\text{ts}}}, \tag{4.9}$$

which can also be represented in terms of y_{d_s} as

$$\tau_{d_s} = \frac{y_{d_s} \prod_{s \in S} \prod_{d'_s \in \mathcal{D}_s, d'_s \neq d_s} (1 + y_{d'_s})}{\prod_{s \in S} \prod_{d_s \in \mathcal{D}_s} (1 + y_{d_s}) - t'}. \tag{4.10}$$

4.3 Problem Formulation

To enable coexistence of different slices in a shared wireless network, an effective slicing of resources is required with two conflicting objectives: maximizing the efficiency and providing isolation between slices. In an unsaturated network, the achievable throughput can be increased by switching between assignment-based and random access-based schemes. As an assignment-based scheme is more efficient for devices with high probabilities of packet transmission, while a random access-based scheme has better performance when devices transmit less frequently. Also, compared to the pure random access-based scheme, splitting devices into two groups, assignment-based and random access-based, leads to a lower number of devices in the random access segment and consequently a smaller number of collisions, and higher utilization. Note that using the pure assignment-based scheme leads to system underutilization as time-slots would be assigned to devices with low traffic demands.

In the proposed reconfigurable access scheme, the scheduling algorithm determines the partition between the DFA and RA segments. More specifically, it determines which devices should transmit in the DFA segment, based on the traffic demand of each device and slice reservations. Furthermore, it derives the parameter p for the rest of devices which compete with each other the in the remaining time of the frame using p-persistent CSMA.

Here, we present the formulation for throughput maximization of this reconfigurable access scheme, assuming that statistical traffic parameters of devices (i.e., a_{d_s}) are known by the AP.

The expected throughput associated with the DFA segment is

$$S_{\text{dfa}}(t) = \sum_{s \in \mathcal{S}} \sum_{d_s \in \mathcal{D}_s} \theta_{d_s}(t)(1 - \psi_{d_s})x_{d_s}(t), \tag{4.11}$$

where $x_{d_s}(t)$ is a binary variable indicating whether a time-slot is allocated to the device d_s in the frame t (i.e., $x_{d_s}(t) = 1$) or not (i.e., $x_{d_s}(t) = 0$) and $X(t) = [x_{d_s}(t)]_{\forall d_s}$. Moreover, the expected throughput of the RA segment can be computed as

$$S_{\text{ra}}(t) = T_{\text{ra}}(t) \sum_{s \in \mathcal{S}} \sum_{d_s \in \mathcal{D}_s} \rho_{d_s}(t), \tag{4.12}$$

where $T_{\text{ra}}(t)$ denotes the duration of the RA segment in the frame t. Taking into account the number of scheduled devices for DFA segment, $T_{\text{ra}}(t)$ can be represented as

$$T_{\text{ra}}(t) = T_f - T_{\text{ts}} \sum_{s \in \mathcal{S}} \sum_{d_s \in \mathcal{D}_s} x_{d_s}(t). \tag{4.13}$$

Furthermore, the instantaneous expected total access airtime for slice s can be obtained as

$$\tau_s(t) = \sum_{d_s \in \mathcal{D}_s} \left[T_{\text{ts}}x_{d_s}(t) + T_{\text{ra}}(t)\tau_{d_s}(t) \right], \tag{4.14}$$

where the first term represents the time assigned to devices belonging to the slice s during the DFA segment and the second term indicates the average total access time of devices of the slice s in the RA segment. Finally, at frame t, the AP should solve the following optimization problem in order to obtain $X(t)$ and $Y(t)$.

$$\max_{X(t), Y(t)} \quad S_{\text{dfa}}(t) + S_{\text{ra}}(t), \tag{4.15}$$

subject to,

C4.15.1: $\tau_s(t) \geq r_s$, $\forall s \in \mathcal{S}$,

C4.15.2: $x_{d_s}(t) p_{d_s}(t) = 0$, $\forall s \in \mathcal{S}, \forall d_s \in \mathcal{D}_s$,

C4.15.3: $T_{ts} \sum_{s \in \mathcal{S}} \sum_{d_s \in \mathcal{D}_s} x_{d_s}(t) \leq T_{\max}$,

C4.15.4: $p_{d_s}(t) \leq 1$, $\forall s \in \mathcal{S}, \forall d_s \in \mathcal{D}_s$,

C4.15.5: $x_{d_s}(t) \in \{0, 1\}$, $\forall s \in \mathcal{S}, \forall d_s \in \mathcal{D}_s$.

In this optimization problem, the objective function represents the total network throughput in both DFA and RA segments for the frame t. The constraint C4.15.1 is to guarantee that the reservation of each slice is met. Moreover, C4.15.2 ensures that the device d_s is only selected for either DFA or RA segment. C4.15.3 limits the number of devices that could transmit in the DFA segment. Finally, C4.15.4 indicates that $p_{d_s}(t)$ should be less than or equal to one and C4.15.5 states that $x_{d_s}(t)$ is a binary variable. In the rest of the chapter, t is omitted in all equations for the sake of simplicity. Substituting (4.11), (4.12), (4.13), and (4.14) in (4.15), the optimization problem can be written as

$$\max_{X,Y} \sum_{s \in \mathcal{S}} \sum_{d_s \in \mathcal{D}_s} (1 - \psi_{d_s}) \left[\theta_{d_s} x_{d_s} + \frac{y_{d_s}(T_f - T_{ts} \sum_{s \in \mathcal{S}} \sum_{d_s \in \mathcal{D}_s} x_{d_s})}{\prod_{s \in \mathcal{S}} \prod_{d_s \in \mathcal{D}_s} (1 + y_{d_s}) - t'} \right], \qquad (4.16)$$

subject to,

C4.16.1: $\sum_{d_s \in \mathcal{D}_s} \left[T_{ts} x_{d_s} + \frac{y_{d_s} \prod_{s \in \mathcal{S}} \prod_{d'_s \in \mathcal{D}_s, \neq d_s} (1 + y_{d'_s})}{\prod_{s \in \mathcal{S}} \prod_{d_s \in \mathcal{D}_s} (1 + y_{d_s}) - t'} \right.$

$\left. (T_f - T_{ts} \sum_{s \in \mathcal{S}} \sum_{d_s \in \mathcal{D}_s} x_{d_s}) \right] \geq r_s$, $\forall s \in \mathcal{S}$,

C4.16.2: $x_{d_s} y_{d_s} = 0$, $\forall s \in \mathcal{S}, \forall d_s \in \mathcal{D}_s$,

C4.16.3: $T_{ts} \sum_{s \in \mathcal{S}} \sum_{d_s \in \mathcal{D}_s} x_{d_s} \leq T_{\max}$,

C4.16.4: $\dfrac{y_{d_s}}{\theta_{d_s}(1 + y_{d_s})} \leq 1$, $\forall s \in \mathcal{S}, \forall d_s \in \mathcal{D}_s$,

C4.16.5: $x_{d_s} \in \{0, 1\}$, $\forall s \in \mathcal{S}, \forall d_s \in \mathcal{D}_s$.

It is clear that the optimization problem in (4.16) has a non-convex objective function due to the couplings in the RA throughput and involves non-linear constraints with the combination of continuous and binary variables, i.e., y_{d_s} ($y_{d_s} \geq 0$) and x_{d_s} ($x_{d_s} \in \{0, 1\}$). Consequently, (4.16) is a non-convex mixed-integer,

NP-hard optimization problem. Therefore, an efficient algorithm with reasonable computational complexity is needed to solve this scheduling problem.

4.4   Reconfigurable Access Scheme Scheduling with Traffic Knowledge

In this section, we first formulate the scheduling problem (4.16) as a CGP and propose an iterative algorithm to solve it. Then, we discuss the computational complexity of the proposed algorithm. After that, for scenarios of large number of devices, we transform the optimization problem by using approximations and solve it using a two-step decomposition method.

4.4.1   Reconfigurable Access Scheme Scheduling via CGP

The formulated problem in (4.16) is non-convex and thus intractable to solve. To reduce the complexity, we relax the binary variable x_{d_s} into a continuous one in the interval of [0,1]. The induced problem potentially looks like a CGP problem. Based on successive convex approximation, a computationally tractable iterative algorithm can be developed to solve a CGP problem. More specifically, a CGP problem can be transformed to a GP by monomial approximations and then a series of GPs can be solved iteratively to obtain the solution.[1]

Here, we describe how to transform the problem (4.16) into a CGP form and then solve it iteratively by applying monomial approximations as discussed in Appendix 1. First, we can maximize the objective function, by minimizing its negative. However, in CGP the objective function should be positive, and this can be done by adding a sufficiently large constant H. Moreover, we introduce three auxiliary variables $u_{d_s} = 1 + y_{d_s}$, $v_0 = \prod_{s \in \mathcal{S}} \prod_{d_s \in \mathcal{D}_s} u_{d_s} - t'$ and $T_{ra} = T_f - T_{ts} \sum_{s \in \mathcal{S}} \sum_{d_s \in \mathcal{D}_s} x_{d_s}$. By replacing these auxiliary variables with their corresponding terms and applying the aforementioned changes in the objective function, the problem becomes

$$\min_{X,Y,U,T_{ra},v_0} H - \sum_{s \in \mathcal{S}} \sum_{d_s \in \mathcal{D}_s} (1 - \psi_{d_s})\left[\theta_{d_s} x_{d_s} + T_{ra} y_{d_s} v_0^{-1}\right], \qquad (4.17)$$

subject to,

---

[1] A brief overview of CGP and monomial approximations is provided in appendix section "Appendix 1: A Brief Overview of Complementary Geometric Programming".

C4.17.1: $\sum_{d_s \in \mathcal{D}_s} \left[T_{ts} x_{d_s} + T_{ra} y_{d_s} v_0^{-1} \prod_{\forall s \in \mathcal{S}} \prod_{d_s' \in \mathcal{D}_s, \neq d_s} u_{d_s'} \right] \geq r_s, \quad \forall s \in \mathcal{S},$

C4.17.2: $x_{d_s} y_{d_s} = 0, \quad \forall s \in \mathcal{S}, \forall d_s \in \mathcal{D}_s,$

C4.17.3: $T_{ts} \sum_{s \in \mathcal{S}} \sum_{d_s \in \mathcal{D}_k} x_{d_s} \leq T_{\max},$

C4.17.4: $u_{d_s} = 1 + y_{d_s}, \quad \forall s \in \mathcal{S}, \forall d_s \in \mathcal{D}_s,$

C4.17.5: $v_0 = \prod_{s \in \mathcal{S}} \prod_{d_s \in \mathcal{D}_k} u_{d_s} - t',$

C4.17.6: $T_{ra} = T_f - T_{ts} \sum_{s \in \mathcal{K}} \sum_{d_s \in \mathcal{D}_s} x_{d_s},$

C4.17.7: $\dfrac{y_{d_s}}{\theta_{d_s} u_{d_s}} \leq 1, \quad \forall s \in \mathcal{S}, \forall d_s \in \mathcal{D}_s,$

C4.17.8: $x_{d_s} \leq 1, \quad \forall s \in \mathcal{S}, \forall d_s \in \mathcal{D}_s.$

In (4.17), the objective function is not posynomial because of the negative multiplicative in the second term. This can be handled by introducing and minimizing a new auxiliary variable x_0 in addition to guaranteeing the following constraint C4.18.7. The resulting optimization problem is

$$\min_{X,Y,U,T_{ra},v_0,x_0} x_0, \qquad\qquad (4.18)$$

subject to,

C4.18.1: $\dfrac{r_s}{\sum\limits_{d_s \in \mathcal{D}_s} \left[T_{ts} x_{d_s} + T_{ra} y_{d_s} v_0^{-1} \prod\limits_{\forall s \in \mathcal{S}} \prod\limits_{d_s' \in \mathcal{D}_s, \neq d_s} u_{d_s'} \right]} \leq 1, \quad \forall s \in \mathcal{S},$

C4.18.2: $\dfrac{1}{1 + x_{d_s} y_{d_s}} = 1, \forall s \in \mathcal{S}, d_s \in \mathcal{D}_s,$

C4.18.3: $T_{ts} T_{\max}^{-1} \sum_{s \in \mathcal{S}} \sum_{d_s \in \mathcal{D}_s} x_{d_s} \leq 1,$

C4.18.4: $\dfrac{u_{d_s}}{1 + y_{d_s}} = 1, \forall s \in \mathcal{S}, d_s \in \mathcal{D}_s,$

C4.18.5: $\dfrac{\prod_{s \in \mathcal{S}} \prod_{d_s \in \mathcal{D}_s} u_{d_s}}{t' + v_0} = 1,$

C4.18.6: $\dfrac{T_f}{T_{ra} + T_{ts} \sum\limits_{s \in \mathcal{S}} \sum\limits_{d_s \in \mathcal{D}_s} x_{d_s}} = 1,$

C4.18.7: $\dfrac{H}{x_0 + \sum\limits_{s \in \mathcal{S}} \sum\limits_{d_s \in \mathcal{D}_s} (1 - \psi_{d_s})[\theta_{d_s} x_{d_s} + T_{ra} y_{d_s} v_0^{-1}]} \leq 1,$

C4.18.8: $\dfrac{y_{d_s}}{\theta_{d_s} u_{d_s}} \leq 1, \qquad \forall s \in \mathcal{S}, \forall d_s \in \mathcal{D}_s,$

C4.18.9: $x_{d_s} \leq 1, \qquad \forall s \in \mathcal{S}, \forall d_s \in \mathcal{D}_s.$

In this optimization problem, all inequality constraints are in the form of a ratio between two posynomials and equality constraints are in the form of a ratio between a monomial and a posynomial, as in a CGP problem. As discussed in appendix section "Appendix 1: A Brief Overview of Complementary Geometric Programming", the algorithm to deal with CGP consists of monomial approximations and solving a sequence of resulting GP problems until convergence happens. The proposed algorithm to solve (4.18) is described in Algorithm 2.

4.4.2 Computational Complexity of the Proposed Algorithm

The computational complexity of Algorithm 2 consists of two parts:

1. Converting the CGP problem to a GP problem by applying AGMA approximations
2. Solving the GP problem.

For the first part, AGMA approximations cost is $10N_d + S + 1$ operations and this means the computational complexity of approximations is $C_{APP} = \mathcal{O}(N_d)$.

For the second part, the GP problem is solved by CVX using the interior point method. According to [12], the number of iterations (Newton steps), required by this iterative method to solve the GP problem is $n_i = \log(n_c/t^0 \rho) \log \varepsilon$, where n_c denotes the total number of constraints, t^0 is the initial point for applying the interior point method, $0 < \rho \ll 1$ is the stopping criterion and ε is used for updating the accuracy of the method. In the optimization problem (4.18), the number of constraints is $n_c = 4N_d + S + 4$. Thus, the required number of iterations is

$$\mathcal{O}(\log N_d). \qquad (4.19)$$

Each iteration (Newton step) of the interior point method costs $\mathcal{O}(n_c n_v^2)$ operations, where n_v denotes the number of variables. For the optimization problem (4.18), $n_v = 3N_d + 3$. Consequently, the computational complexity for solving the GP problem is

$$\mathcal{O}(N_d^3). \qquad (4.20)$$

The total computational complexity for solving each GP problem is equal to the complexity of each iteration (Newton step) multiplied by the required number of iterations to solve the GP problem. Therefore, we have

---

Algorithm 2 Reconfigurable Access Scheme scheduling via CGP

Input: $\Theta, \Psi, T_{\max}, \varsigma, r_s \, \forall s \in \mathcal{S}$
Initialization: Set initial value to $(X, Y, U, T_{\mathrm{ra}}, v_0, x_0)$
repeat
    Step 1: Monomial approximation

1. Compute ζ^p for denominators of C4.18.1 and C4.18.7
2. Use (4.29) to approximate the posynomials
3. Compute ζ^q for denominators of C4.18.2, C4.18.4, C4.18.5, C4.18.6
4. Use (4.30) to approximate the posynomials

    Step 2: Solve the transformed GP problem

1. replace denominators of (4.18) with obtained monomial terms in **Step 1**
2. $(X', Y', U', T'_{\mathrm{ra}}, v'_0, x'_0) \leftarrow$ solve (4.18)

until $|x_0 - x'_0| < \varsigma$
$p_{d_s} \leftarrow \dfrac{y_{d_s}}{\theta_{d_s}(1+y_{d_s})}$
Set $x_{d_s} = 1$ if it is in the sum(X_s) highest value of X, otherwise set $x_{d_s} = 0$
Output: X, P

$$C_{\mathrm{GP}} = \mathcal{O}(N_{\mathrm{d}}^3 \log N_{\mathrm{d}}). \tag{4.21}$$

Having the computational complexity of both parts of Algorithm 2, we can obtain the total computational complexity of this algorithm. As the complexity order of converting the CGP to the GP (C_{APP}) is less than solving the GP problem (C_{GP}), the order of computational complexity for each iteration of Algorithm 2 is equal to C_{GP}.

4.4.3 Scalable Reconfigurable Access Scheme for Dense Networks

Although the CGP-based algorithm has polynomial complexity, for a massive number of devices an algorithm with less computational complexity is needed [13–15]. To this end, assuming $N_{\mathrm{d}} \gg 1$, we first approximate the RA throughput as

$$\rho_{d_s} = \frac{y_{d_s}(1-\psi_{d_s})}{\prod_{s\in\mathcal{S}}\prod_{d_s\in\mathcal{D}_s}(1+y_{d_s})-t'} \approx \frac{y_{d_s}(1-\psi_{d_s})}{\prod_{s\in\mathcal{S}}\prod_{d_s\in\mathcal{D}_s}(1+y_{d_s})}$$

$$\approx y_{d_s}(1-\vartheta_{d_s})(1-\textstyle\sum_{s\in\mathcal{S}}\sum_{d_s\in\mathcal{D}_s}y_{d_s}). \tag{4.22}$$

Moreover, the RA airtime can be approximated as

$$\tau_{d_s} = \frac{y_{d_s}\prod_{s\in\mathcal{S}}\prod_{d'_s\in\mathcal{D}_s,d'_s\neq d_s}(1+y_{d'_s})}{\prod_{s\in\mathcal{S}}\prod_{d_s\in\mathcal{D}_s}(1+y_{d_s})-t'} \approx \frac{y_{d_s}}{1+y_{d_s}}. \tag{4.23}$$

Replacing (4.22) and (4.23) into (4.16), the optimization problem can be expressed as

$$\max_{X,Y} \sum_{s \in \mathcal{S}} \sum_{d_s \in \mathcal{D}_s} (1 - \psi_{d_s}) \left[\theta_{d_s} x_{d_s} + y_{d_s} \left(1 - \sum_{s \in \mathcal{S}} \sum_{d_s \in \mathcal{D}_s} y_{d_s} \right) \right]$$

subject to: $\qquad\qquad\qquad\qquad\qquad\qquad\qquad\qquad\qquad$ (4.24)

C4.24.1: $\sum_{d_s \in \mathcal{D}_s} \left[T_{ts} x_{d_s} + \dfrac{y_{d_s}}{1 + y_{d_s}} \times \right.$

$\qquad\qquad \left. (T_f - T_{ts} \sum_{s \in \mathcal{S}} \sum_{d_s \in \mathcal{D}_s} x_{d_s}) \right] \geq r_s, \ \forall s \in \mathcal{S}$

C4.24.2: $x_{d_s} p_{d_s} = 0, \qquad \forall s \in \mathcal{S}, \forall d_s \in \mathcal{D}_s$

C4.24.3: $T_{ts} \sum_{s \in \mathcal{S}} \sum_{d_s \in \mathcal{D}_s} x_{d_s} \leq T_{max}.$

To solve this optimization problem, we employ an iterative approach, in which each iteration consists of two steps. At each step, we solve the optimization problem over one variable, while for the other variable we use the value obtained from the last iteration. That is, for scalable reconfigurable access scheme, we first maximize over X for fixed Y, then we maximize over Y for fixed X. The iteration continues until convergence happens between the results of two last rounds. In the following, details of this algorithm are presented.

The first step of the algorithm is to solve the optimization problem over X. In fact, for a fixed value of Y, the optimization problem becomes

$$\max_{X} \sum_{s \in \mathcal{S}} \sum_{d_s \in \mathcal{D}_s} (1 - \psi_{d_s}) \left[\theta_{d_s} x_{d_s} + y_{d_s} \left(1 - \sum_{s \in \mathcal{S}} \sum_{d_s \in \mathcal{D}_s} y_{d_s} \right) \right]$$

subject to:

C4.24.1 & C4.24.3 $\qquad\qquad\qquad\qquad\qquad\qquad\qquad\qquad$ (4.25)

Clearly, this optimization problem is linear due to its linear objective function and constraints with respect to X. In the second step of the algorithm, we maximize the optimization problem over Y for a fixed X. Here, the optimization problem is

$$\max_{Y} \sum_{s \in \mathcal{S}} \sum_{d_s \in \mathcal{D}_s} (1 - \psi_{d_s}) y_{d_s} \left(1 - \sum_{s \in \mathcal{S}} \sum_{d_s \in \mathcal{D}_s} y_{d_s} \right)$$

subject to:

C4.24.1

This problem has a concave constraint but non-concave objective function, thus it is a non-convex optimization problem. By doing a simple manipulation, we rewrite the objective function as difference of two concave functions as follows

$$
\max_{Y} \sum_{s \in \mathcal{S}} \sum_{d_s \in \mathcal{D}_s} (1 - \psi_{d_s})
$$

$$
\left[(y_{d_s} - y_{d_s}^2) - \sum_{s' \in \mathcal{S}} \sum_{d_s' \in \mathcal{D}_s, d_s' \neq d_s} \frac{y_{d_s}^2 + y_{d_s}'^2}{2} \right] -
$$

$$
\sum_{s \in \mathcal{S}} \sum_{d_s \in \mathcal{D}_s} \sum_{s' \in \mathcal{S}} \sum_{d_s' \in \mathcal{D}_s, d_s' \neq d_s} -[(1 - \psi_{d_s}) \frac{(y_{d_s} - y_{d_s}')^2}{2}],
$$

subject to: C4.24.1   (4.26)

With this reformulation, the problem falls into the category of difference of convex functions (DC) programming. An overview of DC programming is presented in appendix section "Appendix 2: An Overview of Difference of Convex".

To solve this optimization problem, we use an iterative approach, wherein at each iteration the second term of the objective function is linearized by Taylor expansion. The details of this algorithm can be found in Algorithm 3.

At each iteration, the computational complexity of Algorithm 3 consists of two steps: (1) solving a linear optimization problem and (2) obtaining the DC-programming results. The linear programming can be efficiently solved by using existing methods, since it has only $S + 1$ constraints. The second step, i.e. DC programming is solved in an iterative manner. Each iteration of this algorithm consists of a convex optimization problem, which due to the low number of constraints, i.e. S, can be solved in an efficient manner. Based on the simulation results, the average number of DC iterations does not vary much over number of devices. Furthermore, the outer iteration terminates after a few rounds.

4.5 Illustrative Results

The simulation is done in Matlab and GP problems are solved using CVX [9]. For evaluation, we study the throughput, defined as the number of packets successfully transmitted in a frame (pckt/f), and delay which represents the number of frames between the time that a packet is generated until it is received by the AP. Results are compared with following methods:

- *p-persistent CSMA*: In this scheme, all devices compete with each other by performing *p*-persistent CSMA. Parameter *p* is the same for all devices and it is set to 0.05, as this value is close to the optimal value for this setting.

- *Random Hybrid DFA-RA*: In this scheme, no traffic arrival statistic is taken into account; at each frame, T_{max}/T_{ts} time-slots are assigned to the devices randomly, while the rest of devices compete in the RA segment with $p = 0.05$. The reason that we use this algorithm is to show how considering traffic parameters can enhance the network performance.
- *Distributed queuing (DQ)*: In this scheme as described in Chap. 2, Sect. 2.1.5, a virtual queue is used to store the requests of the devices. Devices access the channel based on their position in the queue [16].

For the parameter setting, we consider a network of an AP, and two slices with all devices within the communication range. The reason to choose two slices is to better demonstrate the dynamics of direct effects of change in one slice on another. We assume that each frame is divided into 16 time-slots and the length of each time-slot, T_{ts}, is equal to 12 time units. The simulation time is set to 100 frames and each simulation is repeated 10 times. We also set the reservation of each slice equal to $r_s = 6$, the maximum length of each queue equal to $Q_{max} = 10$, $T_{max} = 10$, and the convergence parameter $\varsigma = 0.05$. Furthermore, we assume that channel parameters are as follows: path loss exponent $\zeta = 3$, receiver threshold $\xi_R = 0$ dB, and $\frac{P_t}{\sigma_n^2} = 20$ dB.

4.5.1 CGP-Based Reconfigurable Access Scheme

First, we consider a network, where the packet arrival probabilities of two slices are set as $A_1 = \{[0.8]_5, [0.4]_8\}$ and $A_2 = \{[0.8]_5, [0.4]_{N_2-4}\}$. The notation $[f]_{c_f}$ indicates that there are c_f devices having the same packet arrival probability of f. Devices are randomly located in a circular area following a uniform distribution. The radii of the circular areas are 2 m and 5 m for devices with packet arrival probability

---

Algorithm 3 Scalable Reconfigurable MAC

Input: $\Theta, \Psi, T_{max}, r_s \ \forall s \in \mathcal{S}$
Initialization: Set initial value to (X, Y)
repeat
    $(X', Y') \leftarrow (X, Y)$
    Step 1: Find X

-     $X \leftarrow$ Solve the optimization problem (4.25) for fixed Y

    Step 2: Find Y

-     $Y \leftarrow$ Solve the optimization problem (4.25) for fixed X

until $|(X', Y') - (X, Y)| < \varsigma'$
$p_{d_s} \leftarrow \frac{y_{d_s}}{\theta d_s (1 + y_{d_s})}$
Set $x_{d_s} = 1$ if it is in the sum(X_s) highest value of X, otherwise set $x_{d_s} = 0$
Output: X, P

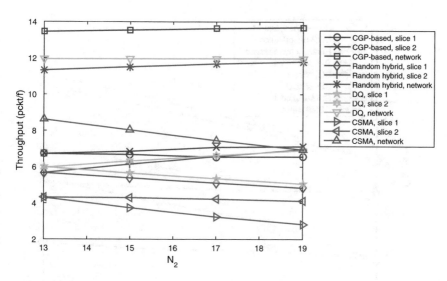

Fig. 4.2 Throughput versus N_2

of 0.8, and the rest, respectively. Furthermore, the parameter C_{dq} of the DQ access scheme is set to 4 time units. To study how well the isolation among slices can be protected in the presence of a variation in one slice, the throughput of both slices is plotted for different numbers of devices in slice 2, while no parameter has changed in slice 1. As shown in Fig. 4.2, by increasing N_2, the throughput of slice 1 degrades slightly, while its reservation is still met. The reason is that larger values of p are assigned to the devices of this slice to keep it isolated from any variation in slice 2. However, the throughput of slice 2 increases since more packets are generated in this slice and therefore assigned time-slots to this slice are left idle with a lower probability. However, by using p-persistent CSMA, DQ, and random hybrid access scheme, the throughput of slice 1 degrades as N_2 increases since the devices have less chance to transmit their packets. The network throughput also decreases for the pure p-persistent scheme due to a larger number of collisions, but remains almost the same for the DQ scheme (since its throughput is mainly dependent on the ratio of the duration of the contention resolution and data transmission phase which is constant for this setup), and the random hybrid scheme (due to the increment in DFA throughput). Although generally increasing the number of devices can affect the successful transmission probability of ARS which consequently impacts the throughput, the effect is not noticeable in this scenario.

Delay results in terms of number of frames are shown in Fig. 4.3. As observed, the CGP-based algorithm almost outperforms other schemes. This is due to the fact that with piggyback mechanism, a device can request for a time-slot. As a result, the probability that a device may have packet for transmission is updated at each frame and based on that free time-slots are assigned to the devices. This prevents starvation or long delay when a device has a packet for transmission. Since in CSMA, devices

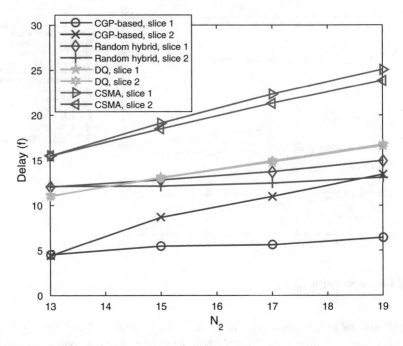

Fig. 4.3 Delay versus N_2

compete with each other and a device having a packet for transmission may fail or does not get a chance to transmit after several time frames. In general, delay increases with increasing N_2. For the other three schemes, average delay is quite similar for both slices 1 and 2. On the other hand, for the proposed CGP-based scheme, a noticeably larger delay increase with N_2 in slice 2 as compared to slice 1. This indicates a much better slice isolation offered by the proposed scheme, i.e., change in slice 2 (i.e., increasing N_2) affects the QoS (i.e., delay) of slice 2 but has a much lower effect on the QoS (i.e., delay) of slice 1.

We also obtain the results for higher numbers of slices. Figure 4.4 shows the average throughput for $S = 2, 4, 6$, and 8. In these simulations, all slices have the same traffic parameters, while the last slice has a larger number of devices than the rest. The traffics parameter setting for different slices is as follows: $A_s = \{[0.8]_{\lfloor 10/S \rfloor}, [0.4]_{\lfloor 16/S \rfloor}\}$, for $s < S$, $S \in \{2, 4, 8\}$. The last slice, i.e., $s = S$ when $S \in \{2, 4, 8\}$ has additional devices compared to the other slices and its traffic parameter setting is $\{[0.8]_{\lfloor 10/S \rfloor + 10 \mod S}, [0.4]_{\lfloor 16/S \rfloor + (16 \mod S) + 6}\}$, where a mod b indicates the remainder of a divided by b. For $S = 6$, the traffic parameters of slices are as follows: $\{[0.8]_1, [0.4]_3\}$ for $s < 5$, and $\{[0.8]_5, [0.4]_7\}$ for $s = 6$. Furthermore, the reservation per slice is set to $r_s = 12/S$. As observed, the results show that the throughput of each slice is equal or greater than its airtime reservation, while other schemes fail to provide the isolation. Furthermore, with increasing S, the total throughput decreases. The reason is that in this scenario, the average traffic of

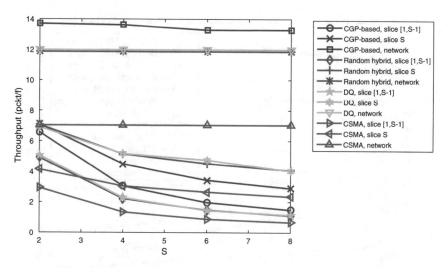

Fig. 4.4 Throughput versus S

each slice is close to its airtime reservation (except the last slice); therefore, to be able to satisfy the airtime reservation of the slice s, a device belonging to this slice with lower expected throughput might be assigned a time-slot, while a device with higher expected throughput from the last slice is not allocated a time-slot.

4.5.2   Scalable Reconfigurable Access Scheme

Here, we consider a network consisting of two slices with traffic parameters as follows: $A_1 = \{[0.95]_5, [0.85]_5, [0.75]_5, [0.65]_5, [0.55]_5, [0.45]_5, [0.35]_5, [0.3]_5, [0.25]_5, [0.15]_5\}$ and $A_2 = \{[0.95]_5, [0.85]_5, [0.75]_5, [0.65]_5, [0.55]_5, [0.45]_5, [0.35]_5, [0.25]_5, [0.15]_5, [0.3]_{N_2-45}\}.$[2] For this scenario, we consider shorter packet sizes each with length of 3 backoff units. Similarly, the length of each time-slot, T_{ts} is 3 backoff units and T_f is 64 time-slots. Furthermore, T_{max} and r_s are set to 50 and 24 time-slots, respectively. Both the inner and outer convergence parameters are set to 0.01. For the DQ, we consider $C_{dq} = 3$ backoff units.

To investigate the algorithm performance in terms of isolation, we fix the number of devices in slice 1 and increase the number of devices of the other slice, N_2. Results in the plotted Fig. 4.5 show that the proposed scalable access scheme outperforms other schemes and slice 1 throughput is almost not affected by increasing N_2. The reason is that the scheduling algorithm takes into account the

---

[2]Such parameters are chosen to include both high load and low load devices as in practice devices with heterogeneous traffic parameters exist.

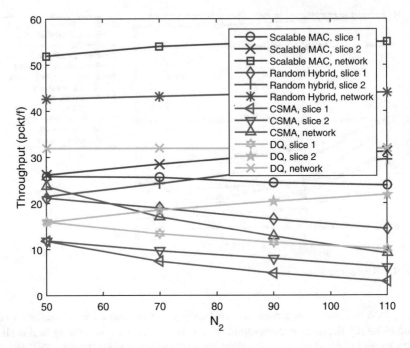

Fig. 4.5 Throughput versus N_2

reservation of each slice, thus isolation is achieved. On the other hand, its throughput slightly increases by increasing N_2 since as the number of packets grows, overall DFA throughput increases while RA throughput is controlled by adjusting the p parameter. The random hybrid and CSMA schemes which employ CSMA with fixed p parameter, thus by increasing N_2, less number of devices of slice 1 get a chance to transmit their packets and furthermore, in case of transmission, there is a higher probability of collision. The reason that the overall throughput of random hybrid access scheme does not drop is that increasing N_2 leads to more congested network meaning that DFA time-slot left idle with lower probability which here almost compensates the CSMA throughput reduction. DQ offers a much lower network throughput than the proposed scheme since for these scenarios devices have small packet sizes and consequently the amount of time devoted for the contention resolution is large compared to the data transmission.

4.6   Concluding Remarks

In this chapter, we have presented a reconfigurable access scheme, where DFA and RA are used for devices with high and low packet transmission probabilities, respectively. This scheduling is formulated as an optimization problem with the objective to maximize the network throughput subject to constraints on slice reservations. To solve this problem, we show that it belongs to the class of CGP, which can be efficiently solved by applying approximations and solving the sequence of resulting GP problems. Furthermore, a scalable algorithm is developed for dense networks. Finally, using simulation results, we show the effectiveness of the proposed algorithms in terms of throughput, isolation and delay.

Appendix 1: A Brief Overview of Complementary Geometric Programming

A geometric programming (GP) is an optimization problem of the form

$$\min_{x} \quad f_0(x) \tag{4.27}$$

subject to:

$$f_i(x) \leq 1, \ i = 1, 2, \ldots, I,$$

$$g_j(x) = 1, \ j = 1, 2, \ldots, J,$$

where $x = [x_1,, \ldots, x_N]$ is a non-negative vector of optimization variables, $g_j(x) = c_j \prod_{n=1}^{N} x_n^{b_{j,n}}$ for all j are monomial functions, and $f_i(x) = \sum_{k=1}^{K_i} c_{i,k} \prod_{n=1}^{N} x_n^{b_{i,k,n}}$ are posynomial functions for $i = 0, \ldots, I$, where coefficients are positive (i.e., $c_j, c_{i,k} > 0$) and $b_{j,n}, b_{i,k,n} \in \mathcal{R}$.

There is a class of optimization problems called complementary geometric programming (CGP), which potentially looks like an extension of GP. In particular, a CGP is presented as

$$\min_{x} \quad P_0(x) \tag{4.28}$$

subject to:

$$P_i(x) \leq 1, \ i = 1, \ldots, I,$$

$$Q_j(x) = 1, \ j = 1, \ldots, J,$$

where $P_0(x)$ is a posynomial and $P_i(x) = \frac{p_i(x)}{p_i^+(x)}$, in which $p_i(x)$ and $p_i^+(x)$ are posynomials. Moreover, $Q_j(x) = \frac{q_j(x)}{q_j^+(x)}$, in which $q_j(x)$ are monomials and $q_j^+(x)$ are posynomials. By approximating $p_i^+(x)$ for all i and $q_j^+(x)$ for all j with monomials, a CGP can be turned into a standard form of GP. Let $p_i^+(x) = \sum_{k=1}^{K_i} h_{i,k}^p(x)$ and $q_j^+(x) = \sum_{k=1}^{K_j} h_{j,k}^q(x)$, where $h_{i,k}^p$ and $h_{j,k}^q$ are monomials. Using AGMA, at iteration l, $p_i^+(x)$ and $q_j^+(x)$ can be approximated as

$$\widetilde{p}_i^+(\boldsymbol{x}(l)) = \prod_{k=1}^{K_i} \left(\frac{h_{i,k}^p(\boldsymbol{x}(l))}{\zeta_{i,k}^p(\boldsymbol{x}(l))} \right)^{\zeta_{i,k}^p(\boldsymbol{x}(l))}, \tag{4.29}$$

$$\widetilde{q}_j^+(\boldsymbol{x}(l)) = \prod_{k=1}^{K_j} \left(\frac{h_{j,k}^q(\boldsymbol{x}(l))}{\zeta_{j,k}^q(\boldsymbol{x}(l))} \right)^{\zeta_{j,k}^q(\boldsymbol{x}(l))}, \tag{4.30}$$

The parameters $\zeta_{i,k}^p(\boldsymbol{x}(l))$ and $\zeta_{j,k}^q(\boldsymbol{x}(l))$ can be computed as

$$\zeta_{i,k}^p(\boldsymbol{x}(l)) = \frac{h_{i,k}^p((\boldsymbol{x}(l-1))}{p_i^+((\boldsymbol{x}(l-1))}, \quad \forall i,\ k, \tag{4.31}$$

$$\zeta_{j,k}^q(\boldsymbol{x}(l)) = \frac{h_{j,k}^q((\boldsymbol{x}(l-1))}{q_j^+((\boldsymbol{x}(l-1))}, \ \forall k,\ j, \tag{4.32}$$

where $\boldsymbol{x}(l-1)$ is the last-round solution of the optimization problem. It is proved that AGMA gives the best local monomial approximation for a posynomial function[12].

Appendix 2: An Overview of Difference of Convex

Difference of convex (DC) functions are formally defined as follows [17, 18].

Let f be a real valued function which maps $\mathcal{R}^n \longrightarrow \mathcal{R}$. We call f a DC function if functions h and g: $\mathcal{R}^n \longrightarrow \mathcal{R}$ exist such that f can be decomposed as the difference between g and h:

$$f(X) = g(X) - h(X) \quad \forall X \in \mathcal{R}^n \tag{4.33}$$

The DC programming (DCP) problem is defined as:

$$\min_{X} \quad f_0(X) \tag{4.34}$$

subject to:

$$f_i(X) \le 0,\ i = 1, 2, \ldots, m,$$

where $f_i : \quad \mathcal{R}^n \longrightarrow \mathcal{R}$ is a differentiable DC function for $i = 1, 2, \ldots, m$.

A function $f : \quad \mathcal{R}^n \longrightarrow \mathcal{R}$ is convex if for every $X_1, X_2 \in \mathcal{R}^n$ and every $\alpha \in [0, 1]$, $f(\alpha X_1 + (1 - \alpha)X_2) \le \alpha f(X_1) + (1 - \alpha)f(X_2)$.

References

1. Y. Mehmood, N. Haider, M. Imran, A. Timm-Giel, M. Guizani, M2M communications in 5G: state-of-the-art architecture, recent advances, and research challenges. IEEE Commun. Mag. **55**(9), 194–201 (2017)
2. M.Z. Shafiq, L. Ji, A.X. Liu, J. Pang, J. Wang, Large scale measurement and characterization of cellular machine-to-machine traffic. IEEE/ACM Trans. Networking **21**(6), 1960–1973 (2013)
3. A. Rajandekar, B. Sikdar, A survey of MAC layer issues and protocols for machine-to-machine communications. IEEE Internet Things J. **2**(2), 175–186 (2015)
4. M. Koseoglu, Lower bounds on the LTE-A average random access delay under massive M2M arrivals. IEEE Trans. Commun. **64**(5), 2104–2115 (2016)
5. Y. Liu, C. Yuen, X. Cao, N.U. Hassan, J. Chen, Design of a scalable hybrid MAC protocol for heterogeneous M2M networks. IEEE Internet Things J. **1**(1), 99–111 (2014)
6. M. Chen, J. Wan, S. Gonzalez, X. Liao, V.C. Leung, A survey of recent developments in home M2M networks. IEEE Commun. Surveys Tuts. **16**(1), 98–114 (2013)
7. A. Dalili Shoaei, M. Derakhshani, S. Parsaeefard, T. Le-Ngoc, Efficient and fair hybrid TDMA-CSMA for virtualized green wireless networks, in *Proceeding of IEEE Vehicle Technical Conference (VTC), Montreal, QC, Canada* (2016)
8. A. Dalili Shoaei, M. Derakhshani, T. Le-Ngoc, Reconfigurable and traffic-aware MAC design for virtualized wireless networks via reinforcement learning. IEEE Trans. Commun. **67**(8), 5490–5505 (2019)
9. M. Grant, S. Boyd, CVX: Matlab software for disciplined convex programming, version 2.1 (2014). http://cvxr.com/cvx
10. O. Dementev, O. Galinina, M. Gerasimenko, T. Tirronen, J. Torsner, S. Andreev, Y. Koucheryavy, Analyzing the overload of 3GPP LTE system by diverse classes of connected-mode MTC devices, in *Internet of Things (WF-IoT), Seoul, South Korea* (2014)
11. G. Bianchi, Performance analysis of the IEEE 802.11 distributed coordination function. IEEE J. Sel. Areas Commun. **18**(3), 535–547 (2000)
12. S. Boyd, L. Vandenberghe, *Convex optimization* (Cambridge University, Cambridge, 2004)
13. Z. Dawy, W. Saad, A. Ghosh, J.G. Andrews, E. Yaacoub, Toward massive machine type cellular communications. IEEE Wireless Commun. **24**(1), 120–128 (2016)
14. C. Bockelmann, N. Pratas, H. Nikopour, K. Au, T. Svensson, C. Stefanovic, P. Popovski, A. Dekorsy, Massive machine-type communications in 5G: physical and MAC-layer solutions. IEEE Commun. Mag. **54**(9), 59–65 (2016)
15. B. Han, H.D. Schotten, *Grouping-based Random Access Collision Control for Massive Machine-type Communication* (IEEE, New York, 2017), pp. 1–7
16. A. Laya, C. Kalalas, F. Vazquez-Gallego, L. Alonso, J. Alonso-Zarate, Goodbye, Aloha!. IEEE Access **4**, 2029–2044 (2016)
17. P.D. Tao, L.T.H. An, Convex analysis approach to DC programming: theory, algorithms and applications. Acta Math. Vietnam. **22**(1), 289–355 (1997)
18. R. Horst, N. Thoai, DC programming: overview. J. Optim. Theory Appl. **103**(1), 1–43 (1999)

Chapter 5
Learning-Based Reconfigurable Access Schemes for Virtualized M2M Networks

5.1 Introduction

In Chap. 4, we have proposed a reconfigurable access scheme for MTC systems in which to maximize the network throughput, the traffic statistics of devices are taken into account. More specifically, in the proposed access scheme, time is divided into frames and each frame is further split into two segment: demand/free assignment (DFA) and random access (RA) segment. The DFA segment is allocated to devices having high packet arrival probabilities and the RA segment is for devices with low packet arrival probabilities. In order to partition these segments, the optimal scheduler described in Chap. 4, Algorithm 2 has been proposed. However, this algorithm requires packet arrival probabilities of devices as input which in reality might be unknown a priori by the AP.

In this chapter, considering the scenario of unknown device traffic statistics, we develop a *Thompson sampling*-based algorithm to learn packet arrival probabilities efficiently [1]. Furthermore, we propose a simple algorithm in which the DFA and RA partition is determined by a threshold. In this algorithm, devices having the expected throughput higher than a certain threshold are considered for DFA, while the rest transmit in the RA segment. In particular, we show that the problem to select devices for the DFA segment can be perfectly matched to a thresholding multi-armed bandit. Thresholding multi-armed bandit (TMAB) is a specific type of combinatorial multi-armed bandits (CMAB), where the learner aims to find the set of arms with the mean rewards exceeding a certain threshold, rather than picking a constant number of arms with the highest mean rewards as in CMABs. In the proposed access scheme design, each arm corresponds to a device, and scheduling a device for DFA in each frame is equivalent to playing an arm. The goal is to find a device-selection policy that maximizes the cumulative throughput over finite frames. Thompson sampling has been shown to perform well for CMABs [2].

© Springer Nature Switzerland AG 2020
T. Le-Ngoc, A. Dalili Shoaei, *Learning-Based Reconfigurable Multiple Access Schemes for Virtualized MTC Networks*, Wireless Networks,
https://doi.org/10.1007/978-3-030-60382-3_5

However, its performance for TMABs has not been investigated. In this chapter, we show that Thompson sampling is also a proper and efficient algorithm for TMABs.

Furthermore, to show the efficacy of Thompson sampling algorithm for thresholding multi-armed bandits, we perform the regret analysis, and we prove that it achieves the optimal regret bound for the stochastic TMABs.

The rest of this chapter is organized as follows. We first present an overview of multi-armed bandit problems and algorithms proposed in the literature to solve these problems in Sect. 5.2. Section 5.3 describes the Thompson sampling-based algorithms for scenarios of unknown packet arrival probabilities. Section 5.4 presents the regret analysis for the proposed Thompson sampling-based approach for thresholding access scheme. Section 5.5 provides simulation results. Finally, Sect. 5.6 concludes this chapter.

5.2   Related Works

In this section, we first provide a brief introduction of multi-armed bandit problems and then we present some algorithms proposed in the literature to solve this problem.

5.2.1   Overview of MAB

Generally, an infinite horizon non-Bayesian multi-armed bandit (MAB) is applied to solve sequential decision making problems with the aim to make a selection among multiple choices, each leading to stochastic rewards with partial knowledge of the system. In a classical MAB setting, there is a system of M arms, each having a Bernouli reward distribution with an unknown mean. At each round, $L < M$ arms are chosen to be played. Let $\boldsymbol{\mu} = (\mu_1, \mu_2, \ldots, \mu_M)$ be the vector of mean rewards of all arms, which is unknown to the player. The goal is to repeatedly play these arms in multiple rounds such that the total expected reward over T time steps is maximized.

The expected reward of each arm is estimated, based on its instantaneous reward observations. The accuracy of this estimation directly depends on the number of times that each arm can be selected. Clearly, for a sufficiently large number of choosing each arm, more precise estimation is obtained. This process to estimate the reward of each arm is referred to as the exploration, which inherently time-consuming and the total expected reward is not maximized. In contrast, when the arms with higher expected rewards are frequently pulled, the expected accumulated reward is increased. This is called exploitation of known arms (i.e., maximizing the rewards). Consequently, there is always a trade-off between the exploration and exploitation.

If the expected reward of each arm is known a priori, the optimal action is to choose the arms with the highest expected reward. For the case of unknown reward, the main question is what would be the best policy to choose an arm. A metric to evaluate a policy is *regret*, which is the difference between the expected reward obtained by always choosing the optimal arms, and that obtained by the selected policy [3].

5.2.2 Existing Works

In the context of MAB problem, throughout the rounds, there is always a trade-off between exploration and exploitation. On one hand, the learner wants to exploit the past observations by selecting seemingly good arms. On the other hand, there is always a possibility that the other arms have been underestimated, which gives the motivation to pick unexplored arms in order to gather more information. To deal with such trade-off, various approaches have been proposed such as upper confidence bound (UCB), in which a deterministic index is assigned to each arm. This index represents the sample mean reward of the arm (exploitation term) plus an exploration term, which gives a higher chance to underexplored arms. For UCB-type algorithms, strong theoretical guarantees on the regret can be proved. For example, in [4], the regret bound has been derived for the classical UCB algorithm. For CMABs, the authors in [5] perform the regret analysis for linear rewards, while nonlinear reward bandit has been studied in [6].

The index-based policies such as UCB are popular for CMABs, where L arms with largest indices would be selected in each round. However, TMABs are sensitive to the exact value of estimated mean reward associated to each arm (not relative to others as in CMABs) since they would be compared with a threshold. Thus, in index-based policies where the exploration term is added to the sample mean reward, the index may become far from the real mean reward.

For thresholding multi-armed bandits, Bayesian inference can be a better approach where the unknown parameter (i.e., mean reward) is drawn from a prior probability distribution, that would be updated at each round after the distribution is sampled. This approach allows exploration by randomly sampling from a distribution, where the observed value may fluctuate from the true value. However, the more frequently distribution is sampled and updated, the more certainly the observed value approaches the true value of unknown parameter. One of the old heuristic algorithm based on Bayesian ideas is Thompson sampling. For a long time, this algorithm was not of interest due to the lack of theoretical analysis. However, it has received significant attention after some recent studies [2, 7]. It has been revealed that TS has an excellent performance with the optimal regret bound and also could be applicable to a wider class of problems [8, 9]. Moreover, [10] has derived the regret analysis for a case where TS is used in CMABs.

The TMAB setting has been studied in [11], where a pure exploration algorithm is proposed. In this work, it is assumed that the threshold is known and the goal of the learner is to correctly identify the arms whose means are over or under the threshold up to a certain precision. This algorithm, due to its pure exploration-based nature, cannot be applied to a situation where the aim of learner is to maximize the cumulative reward.

5.3   Learning-Based Reconfigurable Access Scheme via Thompson Sampling

In this section, we present learning-based reconfigurable access schemes for scenarios of unknown packet arrival probabilities. We consider the same network model, frame structure and traffic models described in Chap. 4, Sect. 4.2.

For these scenarios, the input information for Algorithm 2 is not available. One approach is to apply a simple passive learning that uses the empirical mean of packet arrival probabilities as an estimator. In this algorithm, each time a device sends a packet over DFA, its estimated packet arrival probability is updated. The problem is that the devices having higher empirical mean may obtain a higher chance for DFA transmission than the ones that have smaller empirical mean in the past but may show higher mean in the future. Thus, this approach may lead to a huge performance loss over time.

In other words, if we only rely on the exploitation that uses empirical mean as an estimator, we may take the chance from the high-traffic devices that showed low arrival rates in the past. On the other hand, if we assign time-slots to the devices with low empirical mean (exploration), the performance might be decreased because a device that shows low arrival probability in the past might actually be a low traffic device. Therefore, a proper trade-off between exploration and exploitation is needed. As Thompson sampling is able to provide this balance, we use Thompson sampling indices as the inputs for the Algorithm 2 in which, instead of using empirical means, indices are sampled from Beta distributions with means equal to the empirical mean of the devices.

In the following, we first describe the Thompson sampling for classical CMABs. After that, we provide the details of the proposed Thompson-sampling-based algorithm. Then, we develop a thresholding algorithm for the scheduling, model it as a TMAB, and apply TS for learning.

5.3.1   Thompson Sampling

In order to select the arms, at each round, the TS algorithm assigns a score to each arm. This score is randomly generated based on a prior distribution. One convenient choice of priors for Bernoulli rewards is the Beta distribution, which is a family of continuous probability distributions defined in the interval of [0, 1]. Furthermore, it is the conjugate distribution of the Bernoulli distribution, i.e., assuming Beta distribution as prior, the posterior distribution is also from the same family [2].

The probability distribution function (pdf) of the Beta distribution is denoted by beta(α, β), where $\alpha > 0$ and $\beta > 0$ are the shape parameters. The mean of beta(α, β) is equal to $\frac{\alpha}{\alpha+\beta}$, and the variance is $\frac{\alpha\beta}{(\alpha+\beta)^2(\alpha+\beta+1)}$, indicating that the higher are the α and β, the narrower is the concentration of beta(α, β) around the mean. For Bernoulli rewards, after playing each arm, the shape parameters are updated as follows. If the reward obtained by playing arm i is 1, α is incremented by 1 otherwise we have $\beta = \beta + 1$. In other words, the posterior distribution is simply beta$(\alpha + 1, \beta)$ or beta$(\alpha, \beta + 1)$, depending on whether the reward is 1 or 0, respectively. The Thompson sampling algorithm initially assumes the arm m to have prior beta$(1, 1)$ on μ_m, which is natural because beta$(1, 1)$ is the uniform distribution on $[0, 1]$. At each time step, the TS algorithm samples from these posterior distributions of the μ_m's, and plays the arms which have the L largest scores [8]. Thus, the computational complexity of TS is $\mathcal{O}(M)$.

5.3.2 Thompson Sampling for Reconfigurable Access Scheme

Here, we develop a Thompson sampling-based algorithm for the reconfigurable access scheme. The proposed algorithm helps to learn the unknown packet arrival probabilities. To this end, at each round, Thompson sampling indices of devices are passed as inputs to the optimal scheduler presented in Chap. 4, Sect. 4.4. The optimal scheduler indicates the time-slot allocation for devices, i.e., which arms are chosen to be played. Once the DFA segment terminates, packet arrival probabilities of devices are updated. In the following, we describe how the update process is performed.

In the proposed system model, each device has a queue in which a packet generated at a certain time-slot could be maintained in the queue. Thus, when the device d_s is chosen for transmission it may transmit the packet which was generated in the previous frames. Consequently, this could result in a biased shape parameter update in the Thompson sampling, and therefore, a biased estimation of the packet arrival probabilities over a long run.

To avoid a biased estimation, the proposed algorithm takes advantage of the piggybacked extra bit with any transmission which indicates whether the corresponding device has still any packets in its queue or not (i.e., $q_{d_s} = 1$ or $q_{d_s} = 0$). More specifically, the values of α_{d_s} and β_{d_s} would be updated only when $q_{d_s} = 0$. In other words, whenever a device transmits a packet with $q_{d_s} = 1$, the scheduler keeps assigning time-slots to that device in the subsequent frames until it has no more packets in the queue. At this point, the scheduler updates the values of α_{d_s} and β_{d_s}, where α_{d_s} is increased by the number of packets that are successfully transmitted during the last sequence of device d_s's transmission denoted by $w_{d_s}(t)$ and β_{d_s} is increased by $t - v_{d_s}(t) - w_{d_s}(t)$. Note that the increment in α_{d_s} also contains the successful transmission over RA, in case the first packet of this transmission sequence was started by transmitting a packet over RA. Thus, in this approach, even

---

Algorithm 4 TS algorithm for reconfigurable access scheme

$\quad$ **Initialization**: $\alpha = 1, \beta = 1, t = 1, W = 1, V = 1$
$\quad$ **repeat**
$\qquad$ **Step 1:** Sample $\hat{\Theta}(t) \sim \text{beta}(\alpha, \beta)$
$\qquad$ **Step 2:** Run Algorithm 2 with $\hat{\Theta}$
$\qquad$ **Step 3:** Update α, β
$\qquad$ **if** $x_{d_s}(t) = 1$ & $q_{d_s}(t) = 0$ **then**
$\qquad\quad$ $\alpha_{d_s}(t) = \alpha_{d_s}(t-1) + w_{d_s}(t)$
$\qquad\quad$ $\beta_{d_s}(t) = \beta_{d_s}(t-1) + t - v_{d_s}(t) - w_{d_s}(t)$
$\qquad$ **end if**
$\qquad$ **if** $x_{d_s}(t) = 1$ & $q_{d_s}(t) = 1$ **then**
$\qquad\quad$ $\alpha_{d_s}(t) = \alpha_{d_s}(t-1)$
$\qquad\quad$ $\beta_{d_s}(t) = \beta_{d_s}(t-1)$
$\qquad$ **end if**
$\qquad$ $t = t + 1$
$\quad$ **until** $t < T$

RA observations can be used to update the empirical mean, while, in a pure CSMA scheme, this information cannot be used to update these parameters. Let assume that, in a pure CSMA scheme, the AP updates α_{d_s} whenever it receives a packet, and updates β_{d_s} whenever it does not receive a packet from the device. The problem is that if the AP does not receive a packet, it does not mean that the device did not have a packet for transmission. The device might have a packet for transmission, but it does not get a chance to transmit or its sent packet might be collided. Thus, the shape parameters cannot be updated in a correct manner. The details of the proposed algorithm can be found in Algorithm 4.

Note that for scenarios of time-varying packet arrival probabilities, the calculation of Thompson sampling indices needs modifications since they are randomly chosen from Beta distribution with mean equal to the sample mean of the packet arrival probabilities. As the mean varies over time, the information obtained from previous observations should be carefully used and some perturbation should be introduced to enable the tracking of its variations over time. For example, one approach is to use a weighted averaging method to update the parameters of Beta distribution, in which larger weights are assigned to recent observations. However, it is obvious that the regret occurred under time-varying scenarios is larger than invariant scenarios.

5.3.3   *Thompson Sampling for Thresholding Multi-Armed Bandits*

In the proposed reconfigurable access scheme, the aim is to maximize the expected network throughput. To reach this goal, it separates devices into two groups, by allocating devices with higher expected throughput to the DFA segment and the rest to the RA segment. Another interpretation is that devices with expected throughput of larger than a certain threshold are proper candidates for DFA, while it is more efficient not to assign any time-slots for the rest. In the following proposition, we analytically prove this fact. But, first, we define $\vartheta_{d_s} = (1-\psi_{d_s})\theta_{d_s}$, which represents the expected throughput of device d_s, when it is selected for DFA.

Proposition 5.1 *If X^* is the solution of the optimization problem (4.15), then*

$$\vartheta_{d_s} > \gamma^*, \quad \forall d_s \in \mathcal{D}_{dfa}. \tag{5.1}$$

when all devices have distinct ϑ_{d_s} and there is only one slice. In (5.1), $\gamma^* = \max\{\vartheta_{d_s}\}$ for all $d_s \in \mathcal{D}_{ra}$, $\mathcal{D}_{ra}$ is the set of all devices assigned to the RA segment (i.e., $\forall s \in \mathcal{S}, \forall d_s \in \mathcal{D}_s : x^*_{d_s} = 0$), and $\mathcal{D}_{dfa}$ is the set of all devices assigned to the DFA segment (i.e., $\forall d_s : x^*_{d_s} = 1$).

Proof The proof is done by contradiction.
Assume that

$$\exists d' : x^*_{d'} = 1 \text{ and } \vartheta_{d'} < \gamma^*. \tag{5.2}$$

Also, let X' be a sub-optimal solution for the optimization problem in which $x'_{d_s} = x^*_{d_s}$, $\forall d_s \in \mathcal{D}_s$, $\forall s \in \mathcal{S}$, except for $d_s = d'$ and $\text{argmax}(\vartheta_{d_s})$ for all $d_s \in \mathcal{D}_{ra}$. Therefore, denoting $\hat{d} = \text{argmax}(\vartheta_{d_s})$ for all $d_s \in \mathcal{D}_{ra}$, it is clear that $x'_{d'} = 0$ and $x'_{\hat{d}} = 1$, while $x^*_{d'} = 1$ and $x^*_{\hat{d}} = 0$. Thus, we have $X^*\Theta^\mathsf{T} < X'\Theta^\mathsf{T}$, since $\vartheta_{d'} < \vartheta_{\hat{d}} = \gamma^*$ according to (5.2). On the other hand, RA throughput is dependent on the length of the RA segment and the number of devices competing in this segment. Since these two parameters are the same in the optimal and suboptimal solutions, the RA throughput stays the same for these solutions. In other words, the same throughput can be achieved by setting $y'_{d'} = y^*_{\hat{d}}$. This means that the total throughput of (X', Y') is larger than (X^*, Y^*), which contradicts the fact that (X^*, Y^*) is the optimal solution.

Remark 5.1 For multiple slices, the proposed scheduling is like having multiple TMABs on different slices with distinct thresholds.

Remark 5.2 Considering cases where some devices have the same ϑ_{d_s}, Proposition 5.1 still holds when devices have the same expectations larger or smaller than $\gamma*$, which is the breaking point.

Based on Proposition 5.1, an alternative algorithm for the reconfigurable access scheme is a threshold-based algorithm in which devices having expected throughput larger than a certain threshold are chosen for DFA, while the rest are considered for RA. Assuming that packet arrival probabilities are unknown, this problem can be modeled as a TMAB wherein each arm corresponds to a device.

Thresholding multi-armed bandit is a specific class of CMABs, in which the arm is worth playing if its expected reward is larger than a certain threshold (denoted by γ). As a result, when the player has an option to choose from M arms, the

---

Algorithm 5 TS-TMAB algorithm for thresholding reconfigurable access scheme

---

Initialization: $\alpha = 1$, $\beta = 1$
for $t = 1 : T$ **do**
  for arm $m = 1 : M$ **do**
    sample $\hat{\theta}_m(t) \sim$ beta(α_m, β_m)
  end for
  Set $\mathcal{M}(t) = \{m \mid \hat{\theta}_m(t) > \gamma\}$
  if Size of $\mathcal{M}(t)$ is $> \frac{T_{\max}}{T_{\text{ts}}}$ **then**
    Keep only arms with $\frac{T_{\max}}{T_{\text{ts}}}$ largest values of $\hat{\theta}_m(t)$
  end if
  for $m \in \mathcal{M}(t)$ **do**
    Play arm m
    Update α_m and β_m
  end for
end for

---

optimal number of arms which should be played is dependent on the number of arms for which we have $\mu_m > \gamma$. Then, since the expected rewards of different arms are unknown, the number of arms which gives the highest expected reward is not known either. Therefore, index-based policies which choose the L arms with the highest indices are not applicable for TMABs. However, in the TS approach, the scores are randomized around the estimated mean of the arms, thus they are more proper to be compared against the threshold. The TS-TMAB algorithm is presented in Algorithm 5.

5.4 Regret Analysis

In this section, we study the regret bound in the DFA segment of the thresholding reconfigurable access scheme, when the network consists of one slice, $Q_{\max} = 0$ (i.e., there is no queue to store the packets) and $\Gamma = T_{\max}/T_{\text{ts}}$, where Γ indicates the number of optimal arms.

Notations $1\{\mathcal{A}\}$ is an indicator function which is equal to 1 if event $\mathcal{A}$ holds and 0 otherwise. $d(p, q) = p \log(p/q) + (1 - p) \log((1 - p)/(1 - q))$ is the Kullback-Leibler divergence between two Bernoulli distributions with means p and q.

For the TS-TMAB algorithm as explained in Algorithm 5, in Theorem 5.1, we prove that the TS-TMAB algorithm for binary rewards achieves an optimal regret bound.

Theorem 5.1 *The regret of TS-TMAB Algorithm is upper bounded by*

$$R_{\text{reg}}(T) \leqslant \sum_{m \in \Gamma^-} \frac{\Delta_m}{d(\mu_m^+, \gamma^-)} \log(T) + \mathcal{O}\left(\frac{1}{\delta^2}\right), \tag{5.3}$$

where Δ_m represents the regret caused by playing suboptimal arm m and can be upperbounded by $\max_{i \in M} \mu_i - \mu_m$. Also, $\mu_m^+ = \mu_m + \delta$, $\gamma^- = \gamma - \delta$ for $\delta > 0$ and Γ^- is the set of arms for which $\mu_m < \gamma$.

Proof Let first define arm m as suboptimal, if $\mu_m < \gamma$. Different from CMABs, the regret of a TMAB is not only dependent on the number of times that a suboptimal arm is chosen. It is also affected by the number of times that only a sub-set of all optimal arms is played. Suppose $\Gamma^+ = \{m \mid \mu_m > \gamma\}$ as the set of all optimal arms and Γ as the number of optimal arms in Γ^+.

Lemma 5.1 *The regret of TS-TMAB algorithm can be decomposed as*

$$R_{\text{reg}}(T) = R_{\text{reg}}^u(T) + \sum_{m \in \Gamma^-} R_{\text{reg}}^m(T), \tag{5.4}$$

where $R_{\text{reg}}^u(T)$ represents the regret caused when the number of optimal arms played is less than Γ and $R_{\text{reg}}^m(T)$ indicates the regret caused by playing the suboptimal arm $m \in \Gamma^-$.

According to Lemma 5.1, in order to derive an upper bound for R_{reg}, we could separately study regret bounds for R_{reg}^u and R_{reg}^m.

To find an upper bound for R_{reg}^u, let us first define an event $\mathcal{U}(t) = \{\hat{\theta}^* > \gamma^-\}$, where $\hat{\theta}^*$ represents the Γ-th largest element of the vector $\hat{\boldsymbol{\theta}}$. The complement of this event $\neg\mathcal{U}(t)$ represents the situation that the number of selected arms is less than Γ, which implies that at least one of the optimal arms is underestimated and not played.

The regret caused by event $\neg\mathcal{U}(t)$ depends on the number of optimal arms, which are not played, as well as which ones exactly. The worst-case scenario is when no optimal arm is selected. For this case, the regret is upper bounded by Γ, since for each arm m we have $\mu_m \leq 1$. Having an upper bound for the instantaneous event of $\neg\mathcal{U}(t)$, in the next lemma, we calculate an upper bound on the number of occurrences of $\neg\mathcal{U}(t)$ over T, aiming to find $R_{\text{reg}}^u(T)$.

Lemma 5.2 *The regret $R_{\text{reg}}^u(T)$ is upper bounded by*

$$R_{\text{reg}}^u(T) \leq \Gamma \sum_{t=1}^{T} \mathbf{1}\{\neg\mathcal{U}(t)\} \leq \mathcal{O}\left(\frac{1}{(\gamma - \gamma^-)^2}\right) = \mathcal{O}\left(\frac{1}{\delta^2}\right). \tag{5.5}$$

Proof The proof is provided in [10].

Here, we continue by studying the regret incurred by choosing a sub-optimal arm $m \in \Gamma^-$. More specifically, Lemma 5.3 presents an upper bound on $R_{\text{reg}}^m(T)$.

Lemma 5.3 *The regret of each suboptimal arm m is upper bounded as*

$$R^m_{\text{reg}}(T) \leq \frac{\Delta_m}{d(\mu^+_m, \gamma^-)} \log(T) + \mathcal{O}\left(\frac{1}{\delta^2}\right). \qquad (5.6)$$

Proof See Appendix.

According to Lemmas 5.1–5.3, the regret bound in Theorem 5.1 can be concluded.

According to Proposition 5.1 and Theorem 5.1, in the following corollaries, we discuss the regret bounds that can be achieved by applying the proposed TS-TMAB-based algorithm for S slices.

Corollary 5.1 *Considering that there are S different slices, according to Remark 5.2, the reconfigurable access scheme behaves like multiple TMABs with distinct thresholds. Thus, the regret of this problem also can be upperbounded by the aggregate regrets of each TMAB as in the worst case.*

5.5   Illustrative Results

In this section, we present simulation results to evaluate the performance of proposed schemes. All simulations are done in Matlab and GP problems are solved using CVX [12]. The performance metrics that we use are throughput, defined as the number of packets successfully transmitted in a frame (pckt/f), and delay which represents the number of frames between the time that a packet is generated until it is received by the AP. Furthermore, we compare the results with three schemes: the pure p-persistent CSMA, random hybrid DA-RA and DQ. The description of these schemes are provided in Chap. 4, Sect. 4.5.

For the parameter setting, we consider a network of an AP with two slices in which all devices are within the communication range of each other. We assume that each frame consists of 16 time-slots and the length of each time-slot, T_{ts}, is equal to 12 time units. The simulation time is set to 100 frames and each simulation is repeated 10 times. We also set the reservation of each slice equal to $r_s = 6$, $T_{\text{max}} = 10$, and the convergence parameter $\varsigma = 0.05$.

5.5.1   Reconfigurable Access Scheme: Unknown Statistics

Here, we focus on the performance of TS algorithms. In the following, we show the results for different scenarios.

5.5.1.1   Effect of Suboptimal Arms Statistic

Here, we consider a scenario in which packet arrival probabilities of suboptimal arms, i.e., CSMA devices, are increased. The motivation behind defining this

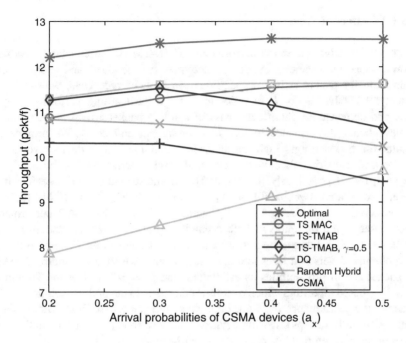

Fig. 5.1 Throughput versus a_x

scenario is to show how the performance of the TS-TMAB is dependent on the closeness of mean reward of suboptimal arms to the optimal arms. In Fig. 5.1, the simulation results are obtained for $T = 100$ frames, where we set $A_1 = \{[0.9]_3, [0.8]_2, [a_x]_{15}\}$, $A_2 = \{[0.95]_3, [0.85]_2, [a_x]_{15}\}$ and a_x represents the packet arrival probabilities of low traffic devices, i.e., suboptimal arms. Furthermore, p is set to 0.05 for all CSMA devices. As observed, the TS-TMAB algorithm with $\gamma = \gamma^*$ achieves better performance compared to the other schemes except the optimal algorithm. The reason is that, for $\gamma = \gamma^*$, although arrival probabilities are unknown, as γ is set to the optimal value, the performance of this scheme becomes closer to the optimal algorithm. However, the performance of the algorithm for $\gamma = 0.5$ drops. Because, for the lower threshold, the probability that CSMA devices are chosen for DFA increases especially for larger a_x. Therefore, for $\gamma = 0.5$, the TS-TMAB scheme has the lowest performance at $a_x = 0.5$. Furthermore, for $a_x = 0.2$, TS access scheme has the largest regret. The reason is that in this case suboptimal devices have larger throughput difference with optimal devices. Therefore, in case they are chosen for DFA, lower throughput will be achieved, leading to the larger regret. Furthermore, it is shown that the throughput of the random hybrid scheme increases by increasing a_x, since time-slots left with a lower probability. For the CSMA and DQ schemes, increasing a_x leads to throughput decrement since the number of collisions increases.

5.5.1.2   Effect of Time

Here, we investigate how the performance of TS-TMAB varies over T. As time grows, more observations on device activities are obtained and more precise estimation for packet arrival probabilities can be achieved. Thus, we obtain the numerical results versus T in order to observe the learning performance over time and demonstrate the effectiveness of learning packet arrival probabilities to achieve better performance in terms of throughput and delay. We provide the results for two settings: high-traffic devices and low-traffic devices versus T. In the first scenario, all devices have high packet arrival probabilities; $A_1 = A_2 = \{[0.9]_2, [0.8]_2, [0.6]_2, [0.55]_2, [0.5]_3\}$. In the second scenario, devices have lower packet arrival probabilities; $A_1 = A_2 = \{[0.8]_2, [0.7]_2, [0.6]_2, [0.2]_{12}\}$. As observed in Figs. 5.2 and 5.3, by increasing T, the TS-TMAB performance gets closer to the optimum in both scenarios. However, for the random hybrid, CSMA and DQ schemes, the performance is not dependent on T. The results are also compared versus a thresholding-based scheme which uses empirical means of packet arrival probabilities as estimators, and $\gamma = \gamma^*$. Unlike the Thompson sampling-based algorithm which over time converges to the optimal solution, in this scheme, the performance does not improve significantly over time. The reason is that the algorithm keeps choosing devices that show good performance in the early steps of the algorithm, while they may have low mean rewards.

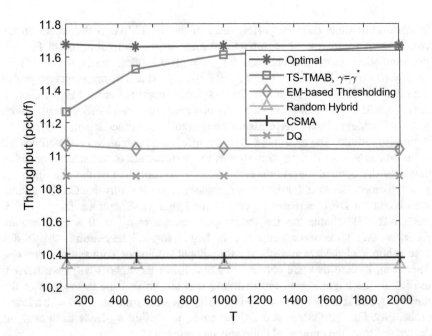

Fig. 5.2 Throughput versus T (high load)

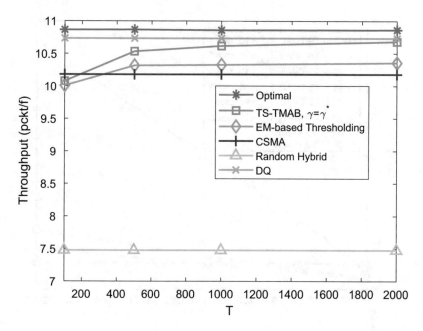

Fig. 5.3 Throughput versus T (low load)

We further obtain the delay results for these two scenarios, where $Q_{max} = 4$, shown in Figs. 5.4 and 5.5. In these figures, we omit the results for CSMA and random hybrid schemes due to their large delay. As observed the delay of TS-TMAB depends on the size of T_{max}. For these parameter settings, larger T_{max} leads to a lower delay. Since at each frame, more devices are scheduled in the DFA, more packets with shorter delay are transmitted. Furthermore, DQ achieves low delay since it applies a queuing strategy which can prevent packets from experiencing long delays.

5.6   Concluding Remarks

In order to deploy the reconfigurable access scheme proposed in Chap. 4, packet arrival statistics are needed by the AP. However, in practice this information may not be known in prior. In this chapter, for these scenarios, two Thompson sampling-based algorithms are proposed. In the first approach, packet arrival probabilities are estimated by Thompson sampling indices, passed to the optimal scheduler to determine which devices should transmit in the DFA segment and which devices can access the the channel in the RA segment. In the second proposed scheme, the reconfigurable access scheme is modeled as a thresholding MAB and a Thompson sampling-based algorithm is proposed to solve that. Furthermore, the regret analysis

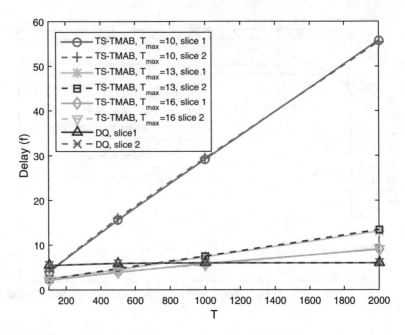

Fig. 5.4 Delay versus T (high load)

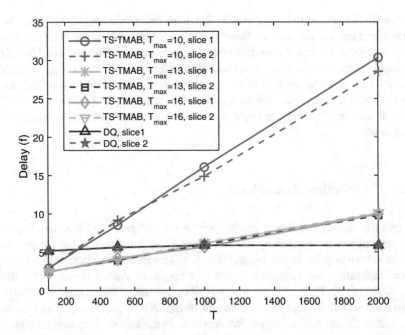

Fig. 5.5 Delay versus T (low load)

is provided for performance evaluation of TS-TMAB algorithm. Finally, using simulation results, we show the effectiveness of the proposed algorithms for unknown packet arrival statistics.

Appendix

Proof of Lemma 5.3

The regret incurred by playing the suboptimal arm m over T is

$$R_{\text{reg}}^m(T) = \sum_{t=1}^{T} \mathbf{1}\{\hat{\theta}_m(t) > \gamma\}\Delta_m. \tag{5.7}$$

To present an upper bound for $R_{\text{reg}}^m(T)$, let us first decompose the event $\mathcal{A}_m = \{\hat{\theta}_m(t) > \gamma\}$ into two complementary sub-events $\mathcal{B}_m = \{\hat{\theta}_m(t) > \gamma, \hat{\mu}_m(t) > \mu_m^-\}$ and $\{\mathcal{C}_m = \varphi_m(t) > \gamma, \hat{\mu}_m \leq \mu_m^-\}$, where $\hat{\mu}_m$ is the empirical mean of arm m. Thus, $R_{\text{reg}}^m(T)$ can be found as

$$R_{\text{reg}}^m(T) = \sum_{t=1}^{T} \mathbf{1}\{\mathcal{B}_m\}\Delta_m + \sum_{t=1}^{T} \mathbf{1}\{\mathcal{C}_m\}\Delta_m \tag{5.8}$$

Here, we calculate the regret bounds for both $\mathcal{B}_m$ and $\mathcal{C}_m$. $\sum_{t=1}^{T} \mathbf{1}\{\mathcal{B}_m\}$ can be bounded as

$$\sum_{t=1}^{T} \mathbf{1}\{\mathcal{B}_m\} \leq \sum_{t=1}^{T} \mathbf{1}\{\hat{\mu}_m(t) > \mu_m^-\} \tag{5.9}$$

According to [9], we have

$$\sum_{t=1}^{T} \mathbf{1}\{\hat{\mu}_m(t) > \mu_m^-\} \leq 1 + \frac{1}{d(\mu_m, \mu_m^-)} = \mathcal{O}\left(\frac{1}{\delta^2}\right) \tag{5.10}$$

As a result, from (5.9) and (5.10), we have

$$\sum_{t=1}^{T} \mathbf{1}\{\mathcal{B}_m\} \leq \mathcal{O}\left(\frac{1}{\delta^2}\right) \tag{5.11}$$

To calculate $\sum_{t=1}^{T} \mathbf{1}\{\mathcal{C}_m\}$, we first define $N_m^{\text{suf}}(T) = \log(T)/d(\mu_m^+, \gamma^-)$, which intuitively is the sufficient number of explorations to make sure that arm m is not worth playing. Then, we continue by decomposing $\mathcal{C}_m$ into two complementary sub-events, $\mathcal{D}_m = \{\hat{\theta}_m(t) > \gamma, \hat{\mu}_m \leq \mu_m^-, N_m(t) \leq N_m^{\text{suf}}(T)\}$ and $\mathcal{E}_m = \{\hat{\theta}_m(t) > \gamma, \hat{\mu}_m \leq \mu_m^-, N_m(t) > N_m^{\text{suf}}(T)\}$, where $N_m(t)$ represents the number times arm m has been played until round t. Consequently,

$$\sum_{t=1}^{T} \mathbf{1}\{\mathcal{C}_m\} = \sum_{t=1}^{T} \mathbf{1}\{\mathcal{D}_m\} + \sum_{t=1}^{T} \mathbf{1}\{\mathcal{E}_m\} \tag{5.12}$$

Simply, $\sum_{t=1}^{T} \mathbf{1}\{\mathcal{D}_m\}$ can be upper bounded by

$$\sum_{t=1}^{T} \mathbf{1}\{\mathcal{D}_m\} \le N_m^{\text{suf}}(T) = \frac{\log(T)}{d(\mu_m^+, \gamma^-)} \tag{5.13}$$

Then, for $\sum_{t=1}^{T} \mathbf{1}\{\mathcal{E}_m\}$, since $N_m(t) > N_m^{\text{suf}}(T)$, we have

$$\sum_{t=1}^{T} \mathbf{1}\{\mathcal{E}_m\} \le \sum_{t=1}^{T} \sum_{n=N_m^{\text{suf}}(T)+1}^{T} \mathbb{P}(\hat{\theta}_m(t) > \gamma \,|\hat{\mu}_m(t) \le \mu_m^-, N_m(t) = n) \tag{5.14}$$

When $N_m(t) = n$, $\hat{\theta}_m$ is sampled from beta$(\hat{\mu}_m(t)(N_m^{\text{suf}}(T) + 1), (1 - \hat{\mu}_m(t))(N_m^{\text{suf}}(T) + 1))$. Considering this fact, based on the Chernoff–Hoeffding inequality bound, it has been proved that

$$\mathbb{P}(\hat{\theta}_m(t) > \gamma \,|\hat{\mu}_m(t) \le \mu_m^-, N_m(t) = n) \le e^{-d(\gamma, \mu_m^-)n} \tag{5.15}$$

Consequently, $\sum_{t=1}^{T} \mathbf{1}\{\mathcal{E}_m\} \le \sum_{t=1}^{T} \sum_{n=N_m^{\text{suf}}(T)+1}^{T} e^{-d(\gamma, \mu_m^-)n}$. By Chernoff bound and Pinsker's inequality, it can be shown that $\sum_{n=N_m^{\text{suf}}(T)+1}^{T} e^{-d(\gamma, \mu_m^-)n}$ is an order of $\mathcal{O}(1/T)$. Subsequently,

$$\sum_{t=1}^{T} \mathbf{1}\{\mathcal{E}_m\} = \sum_{t=1}^{T} \mathcal{O}(1/T) = \mathcal{O}(1). \tag{5.16}$$

Finally, considering (5.11), (5.13), and (5.16), it can be concluded that

$$R_{\text{reg}}^m(T) = (\sum_{t=1}^{T} \mathbf{1}\{\mathcal{B}_m\} + \sum_{t=1}^{T} \mathbf{1}\{\mathcal{D}_m\} + \sum_{t=1}^{T} \mathbf{1}\{\mathcal{E}_m\})\Delta_m \le \frac{\Delta_m \log(T)}{d(\mu_m^+, \gamma^-)} + \mathcal{O}\left(\frac{1}{\delta^2}\right). \tag{5.17}$$

References

1. A. Dalili Shoaei, M. Derakhshani, T. Le-Ngoc, Reconfigurable and traffic-aware MAC design for virtualized wireless networks via reinforcement learning. IEEE Trans. Commun. **67**(8), 5490–5505 (2019)
2. O. Chapelle, L. Li, An empirical evaluation of Thompson sampling, in *Advances in Neural Information Process Systems* (2011), pp. 2249–2257
3. A. Dalili Shoaei, M. Derakhshani, S. Parsaeefard, T. Le-Ngoc, Learning-based hybrid TDMA-CSMA MAC protocol for virtualized 802.11 WLANs, in *Proceedings of the IEEE*

International Symposium on Personal, Indoor and Mobile Radio Communication (PIMRC) (IEEE, Piscataway, 2015), pp. 1861–1866

4. P. Auer, N. Cesa-Bianchi, P. Fischer, Finite-time analysis of the multiarmed bandit problem. Mach. Learn. **47**(2–3), 235–256 (2002)

5. Y. Gai, B. Krishnamachari, R. Jain, Combinatorial network optimization with unknown variables: multi-armed bandits with linear rewards and individual observations. IEEE/ACM Trans. Netw. **20**(5), 1466–1478 (2012)

6. W. Chen, Y. Wang, Y. Yuan, Combinatorial multi-armed bandit: general framework, results and applications, in *Proceedings of the International Conference on Machine Learning* (2013), pp. 151–159

7. S.L. Scott, A modern Bayesian look at the multi-armed bandit. Appl. Stochastic Models Bus. Ind. **26**(6), 639–658 (2010)

8. S. Agrawal, N. Goyal, Analysis of Thompson sampling for the multi-armed bandit problem, in *COLT* (2012), pp. 1–39

9. S. Agrawal, N. Goyal, Further optimal regret bounds for Thompson sampling, in *Artificial Intelligence and Statistics* (2013), pp. 99–107

10. J. Komiyama, J. Honda, H. Nakagawa, Optimal regret analysis of Thompson sampling in stochastic multi-armed bandit problem with multiple plays, in *International Conference on Machine Learning (ICML)*, Lille (2015)

11. A. Locatelli, M. Gutzeit, A. Carpentier, An optimal algorithm for the thresholding bandit problem (2016). arXiv preprint arXiv:1605.08671

12. M. Grant, S. Boyd, CVX: Matlab software for disciplined convex programming, version 2.1 (2014). http://cvxr.com/cvx

Chapter 6
Efficient and Fair Access Scheme for MTC: LTE/WiFi Coexistence Case

6.1 Introduction

In Chaps. 3–5, we have proposed MTC access schemes for scenarios that all devices are connected to the same access network. In this chapter, we consider the case that devices belong to two different networks; some are connected to the LTE, while the rest are WiFi devices [1, 2]. Traditionally, unlicensed bands were only used by some technologies such as WiFi, however to meet the growing wireless data demand and to improve the spectrum efficiency, LTE operation on unlicensed bands has been proposed by the 3GPP, which is the topic of this chapter.

As mentioned above, this chapter deals with the scenario that both WiFi and LTE systems transmit over the unlicensed bands. In this situation, the main concern is that the performance of WiFi devices could be negatively impacted by LTE. In particular, since in LTE a scheduling-based channel access is used, starvation may occur for WiFi devices. The reason is that to transmit a packet in WiFi, the device waits until the channel becomes idle, while LTE adopts an aggressive approach to access the channel [3–8].

In the literature, this issue has received a lot of attention, where most of solutions are for uncoordinated scenarios, meaning that no coordination between the LTE and WiFi exists. Furthermore, most of them address the case where unlicensed bands are used for LTE downlink transmissions. In this chapter, we propose a coordinated structure, where the central network entity facilitates the separation of LTE and WiFi transmissions in two different segments. With no major MAC modification required, LTE devices can access the channel in the schedule-based manner, while in the second segment, WiFi devices can opportunistically transmit their packets using p-persistent CSMA. Moreover, such network entity can enable dynamic scheduling by assigning time-slots to LTE devices and adjusting p for WiFi devices. This can improve the network throughput and preserve the WiFi throughput requirement.

© Springer Nature Switzerland AG 2020

T. Le-Ngoc, A. Dalili Shoaei, *Learning-Based Reconfigurable Multiple Access Schemes for Virtualized MTC Networks*, Wireless Networks, https://doi.org/10.1007/978-3-030-60382-3_6

Furthermore, using such an approach, the unlicensed band can be used for both uplink and downlink LTE transmissions.

Assuming an unsaturated network for both LTE and WiFi systems, the goal of this chapter is to maximize the overall throughput over each duty cycle, while the WiFi throughput does not fall below a target threshold. In other words, this scheme acts like a duty-cycle-based approach, where a period of a time-frame with a variable length is assigned exclusively to the LTE system, which cannot be used by WiFi devices. In order to solve this optimization problem, we formulate it as a complementary geometric programming problem, which can be solved by applying an iterative algorithm.

Furthermore, delay analysis is performed for LTE devices to provide an analytical framework that can be used for designing the LTE admission control policy, e.g., a specific number of devices that can be supported, while meeting their QoS constraints. We consider a system of homogeneous LTE devices, in which all devices have the same packet arrival probabilities. For this system, we derive an analytical upper-bound of average delay by modeling the time-slot assignment to the device with a queuing system which consists of two servers. Consequently, the probability mass function of packet delay is derived and average packet delay is calculated.

The rest of this chapter is organized as follows. We first introduce the system model in Sect. 6.2. Section 6.3 presents the problem formulation and transformation of the optimization problem into a CGP form. The performance analysis for LTE devices in terms of average packet delay is presented in Sect. 6.4. Furthermore, Sect. 6.5 presents the simulation results. Finally, we provide some concluding remarks in Sect. 6.6.

6.2 System Model

6.2.1 Coordinated Structure for LTE/WiFi Coexistence

We consider an IEEE 802.11-based WLAN with N_w, $d_w \in \mathcal{D}_w$ devices sharing the channel with an LTE network serving N_l, $d_l \in \mathcal{D}_l$ devices. We assume that the traffic and channel models are as described in Chap. 3, Sect. 3.2, and Chap. 4, Sect. 3.2, respectively. In particular, packets arrive at LTE and WiFi devices with probabilities of a_{d_l} and a_{d_w} at each duty cycle. We assume that the controller is aware of packet arrival probabilities and at each duty cycle, it updates θ_{d_l} and θ_{d_w} denoting the probability of devices d_l and d_w having a packet for transmission. Furthermore, ψ_{d_l} and ψ_{d_w} represent the outage probabilities for devices d_l and d_w.

In order to enable efficient coexistence between these two networks, we assume that there is a virtual network entity for software-defined wireless networking, which can control both the WiFi AP and the LTE BS in a central manner, where devices from different wireless technologies form different slices: LTE slice and WiFi slice.

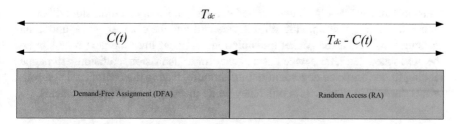

Fig. 6.1 Duty-cycle-based structure for LTE and WiFi coexistence

In the virtualized scheme, devices of different systems access through their distributed access points which are controlled by a central entity. More specifically, this architecture supports separating the data plane from the control plane. The control plane responsible for managing resources is centralized while separated data planes are considered, i.e., data are being forwarded independently from LTE BS and WiFi AP [9].

We use the duty-cycle-based approach in order to access the channel as shown in Fig. 6.1. Here we assume that each duty cycle contains a time-frame similar to the one proposed in Chap. 4, which is divided into two segments. In the first segment, LTE devices can access the channel in the schedule-based manner, while in the second segment, WiFi devices can opportunistically transmit their packets using p-persistent CSMA. To realize this model, we deploy point coordination function (PCF) mechanism for the WiFi. In this mechanism, each superframe consists of a contention-free period (CFP) followed by a contention period (CP). Thus, by activating the PCF, the CFP can be used by LTE devices, while the CP can be assigned to the WiFi devices [10].

Each duty cycle has a fixed duration of T_{dc}, consisting of time-slots with duration of T_{ts}. Although the length of each duty cycle is fixed, the length of each segment assigned to WiFi or LTE varies over different cycles. The duration of LTE transmission is denoted by $C(t)$ and $T_{dc} - C(t)$ that indicates the length of second segment for WiFi transmissions.

This architecture allows each network to use its current deployed MAC protocol, i.e., LTE devices can transmit in DFA segment, while WiFi devices compete with each other in the RA segment. Thus, the model leads to minimal modification requirement for both systems. It also eliminates the LBT requirement for LTE network. Specifically, it facilitates uplink transmissions for LTE network, since in the uplink scenario, the device has to perform the LBT operation while in downlink only BS listens to the channel before transmission. However, as in this approach, devices of different networks access in separated time, unlicensed band can be also used for LTE uplink transmissions with no need for LBT operation. To emphasize the advantage of employing this structure, consider a scenario using a non-LBT approach without any coordination among LTE and WiFi systems, where at the beginning of each duty cycle, the LTE BS schedules the devices for the uplink transmission. If one or some of these scheduled LTE devices have no packets for transmission considering an unsaturated network, the assigned time-slot will be left empty and during that time the WiFi device will find the channel as idle and it may transmit a packet. If the packet transmission continues until the next time-slot and the next scheduled LTE device has a packet for transmission, collision will happen leading to performance degradation. Furthermore, having the centralized control, it brings the benefit of spectrum efficiency while meeting the WiFi requirements.

6.3 Hybrid Scheduling via CGP for LTE and WiFi Coexistence

6.3.1 Problem Formulation

Consider a system consisting of an LTE BS and a WiFi AP, which are connected to a virtual network entity. To facilitate the coexistence between WiFi and LTE and increase the spectral efficiency, the central controller dynamically divides each duty cycle between two slices. This would be done in a way that the overall expected throughput is maximized, while the throughput of WiFi devices does not degrade significantly compared to the case in which the band is not shared with LTE devices. In particular, this problem can be mathematically expressed as follows,[1]

$$\max_{X(t),Y(t)} \quad S_{\text{dfa}}(t) + S_{\text{ra}}(t), \quad \text{subject to,} \qquad (6.1)$$

$$\text{C6.1.1:} \quad S_{\text{ra}}(t) \geq \eta$$

$$\text{C6.1.2:} \quad T_{\text{ts}} \sum_{d_1 \in \mathcal{D}_1} x_{d_1}(t) \geq C_{\min}.$$

Here, $S_{\text{dfa}}(t)$ and $S_{\text{ra}}(t)$ represent the LTE and the WiFi expected throughput at duty cycle t, respectively. $X(t) = [x_{d_1}(t)]$ is the time-slot allocation for LTE devices at duty cycle t. In particular, $x_{d_1}(t) \in \{0, 1\}$, where $x_{d_1}(t) = 1$ if a time-slot is allocated to the LTE device d_1 in the DFA segment, and $x_{d_1}(t) = 0$ otherwise. Furthermore, $Y(t) = [y_{d_w}(t)]$ is the vector, where the element $y_{d_w}(t)$ is as defined in Eq. (4.5). In this optimization problem, the objective function represents the total expected throughput of the network in both DFA and RA segments in the duty cycle t, where the throughput in the DFA and RA segments represents the throughput of the LTE and WiFi systems, respectively. In addition, the first constraint is to guarantee that WiFi throughput does not fall below a required threshold (denoted by η). Furthermore, to keep LTE devices satisfied, we add the second constraint to reserve the airtime with duration of at least $C_{\min}$ for LTE devices. It should be

---

[1]The results can be easily extended for downlink scenario of WiFi as well since the AP acts as a station in downlink and accesses the channel in the same manner.

noted that the values of η and $C_{\min}$ are fixed, dictated based on the service level agreements of WiFi and LTE systems.

Considering the expected throughput of each WiFi device according to Eq. (4.8), the optimization problem in (6.1) can be expanded as[2]

$$\max_{X,Y} \sum_{d_l \in \mathcal{D}_l} (1 - \psi_{d_l}) \theta_{d_l} x_{d_l} + \sum_{d_w \in \mathcal{D}_w} \frac{(1 - \psi_{d_w}) y_{d_w} (T_{dc} - T_{ts} \sum_{d_l \in \mathcal{D}_l} x_{d_l})}{\prod_{d_w \in \mathcal{D}_w} (1 + y_{d_w}) - t'} \tag{6.2}$$

subject to:

$$\text{C6.2.1: } \sum_{d_w \in \mathcal{D}_w} \frac{(1 - \psi_{d_w}) y_{d_w} (T_{dc} - T_{ts} \sum_{d_l \in \mathcal{D}_l} x_{d_l})}{\prod_{d_w \in \mathcal{D}_w} (1 + y_{d_w}) - t'} \geq \eta$$

$$\text{C6.2.2: } T_{ts} \sum_{d_l \in \mathcal{D}_l} x_{d_l} \geq C_{\min}$$

In the objective function of (6.2), the first term represents the expected throughput associated with LTE devices in the DFA segment which is $S_{dfa} = \sum_{d_l \in \mathcal{D}_l} (1 - \psi_{d_l}) \theta_{d_l} x_{d_l}$. Furthermore, the second term denotes the entire WiFi network throughput, i.e., $S_{ra} = T_{ra} \sum_{d_w \in \mathcal{D}_w} \rho_{d_w}$, where T_{ra} denotes the duration of the RA segment in a duty cycle.

Before solving this optimization problem, we provide a discussion on its feasibility. To determine the feasibility of this optimization problem, we obtain the feasible region for C, which indicates the length of the DFA segment in a duty cycle and is the common parameter in the conflicting constraints of C6.2.1 and C6.2.2. According to C6.2.2, we constrain $C \geq C_{\min}$. Furthermore, from C6.2.1, we can obtain the maximum value that C can take, while the WiFi throughput requirement is met. This value is obtained when $S_{ra} = \eta$, however as S_{ra} depends on Y which is the optimization variable of (6.2), derivation of C leading to $S_{ra} = \eta$ is not trivial. Thus, we denote this value of C as a function of η, i.e., $f(\eta)$. Consequently, we have

$$C_{\min} \leq C \leq T_{dc} - f(\eta).$$

This means that as long as $T_{dc} - f(\eta)$ is larger than $C_{\min}$, the optimization problem in (6.2) is feasible. It should be noted that as far as $C_{\min}$ is sufficiently smaller than $T_{dc} - \eta$, the problem would be feasible.

Now that we derived the feasibility region of (6.2), we discuss solving this problem. The optimization problem in (6.2) has a non-convex objective function and a non-convex constraint with the combination of continuous and binary variables. Consequently, (6.2) is a non-convex mixed-integer, NP-hard optimization problem.

---

[2]In the rest of this chapter, t is omitted from all terms.

Therefore, an efficient algorithm with a reasonable computational complexity is needed. In a similar approach as used in Chap. 4 to solve the optimization problem (4.15), we transform the problem into CGP. To do that, first, we maximize the objective function by minimizing its negative. Then, we add a sufficiently large constant H to the objective function, as in CGP the objective function should be positive. Furthermore, we use three auxiliary variables $u_{d_w} = 1 + y_{d_w}$, $T_{ra} = T_{dc} - T_{ts} \sum_{d_l \in \mathcal{D}_l} x_{d_l}$ and $v_0 = \prod_{d_w \in \mathcal{D}_w} u_{d_w} - t'$. By applying all these manipulations to the optimization problem (6.2), we reach to

$$\min_{X,Y,U,T_{ra},v_0} H - \sum_{d_l \in \mathcal{D}_l} (1 - \psi_{d_l}) \theta_{d_l} x_{d_l} - \sum_{d_w \in \mathcal{N}_w} (1 - \psi_{d_w}) T_{ra} y_{d_w} v_0^{-1} \qquad (6.3)$$

subject to:

C6.3.1: $\sum_{d_w \in \mathcal{D}_w} (1 - \psi_{d_w}) T_{ra} y_{d_w} v_0^{-1} \geq \eta$,

C6.3.2: $T_{ts} \sum_{d_l \in \mathcal{D}_l} x_{d_l} \geq C_{min}$,

C6.3.3: $u_{d_w} = 1 + y_{d_w}$, $\forall d_w \in \mathcal{D}_w$,

C6.3.4: $v_0 = \prod_{d_w \in \mathcal{D}_w} u_{d_w} - t'$,

C6.3.5: $T_{ra} = T_{dc} - T_{ts} \sum_{d_l \in \mathcal{D}_l} x_{d_l}$.

The optimization problem (6.3) is still not in the CGP form, as the objective function is not a posynomial due to the negative factor in the second and third terms. To handle this, we reformulate the optimization problem as (6.4) in which the objective function is replaced with a new auxiliary variable x_0 and the constraint C6.4.5 is added.

$$\min_{X,Y,U,T_{ra},v_0,x_0} x_0, \text{ subject to:}$$

C6.4.1: $\dfrac{\eta}{\sum_{d_w \in \mathcal{D}_w} (1 - \psi_{d_w}) T_{ra} y_{d_w} v_0^{-1}} \leq 1$,

C6.4.2: $\dfrac{u_{d_w}}{1 + y_{d_w}} = 1$, $\forall d_w \in \mathcal{D}_w$,

C6.4.3: $\dfrac{\prod_{d_w \in \mathcal{D}_w} u_{d_w}}{t' + v_0} = 1$,

C6.4.4: $\dfrac{T_{dc}}{T_{ra} + T_{ts} \sum_{d_w \in \mathcal{D}_w} x_{d_w}} = 1$,

$$\text{C6.4.5: } \frac{H}{x_0 + \sum_{d_l \in \mathcal{D}_l} (1 - \psi_{d_l})\theta_{d_l} x_{d_l} + \sum_{d_w \in \mathcal{D}_w} (1 - \psi_{d_w})T_{ra} y_{d_w} v_0^{-1}} \leq 1,$$

$$\text{C6.4.6: } \frac{C_{min}}{T_{ts} \sum_{d_l \in \mathcal{D}_l} x_{d_l}} \leq 1.$$

In this optimization problem,[3] all upper-bound inequality constraints are in the form of a ratio between two posynomials and equality constraints are in the form of a ratio between a monomial and a posynomial. Thus, the problem belongs to the class of complementary geometric programming problem.

Using the approximations in Eqs. (4.29) and (4.30), in each iteration, the optimization problem in (6.4) would be in the form of a standard GP problem. Consequently, the optimal solution can be achieved by iteratively applying monomial approximations and solving a series of GPs [11]. The details of this approach are summarized in Algorithm 6.

It should be noted using the algorithm proposed in [12], the P_w value obtained from Algorithm 6 can be implemented in CSMA/CA by configuring the MAC parameters such as minimum contention window, arbitrary inter frame space (AIFS) and retry transmission limit.

6.3.2 Computational Complexity of the Proposed Algorithm

As discussed in Chap. 4, Sect. 4.4.2, the computational complexity to solve each iteration of a CGP problem consists of two steps. The first step is to convert the CGP problem to a GP problem. For optimization problem (6.4), the AGMA approximations require $2N_l + 4N_w + 4$ operations, which its computational complexity is $C_{APP} = \mathcal{O}(N_l + N_w)$. In the second step, the resulted GP is solved by the interior point method which has a computational complexity of $\mathcal{O}(n_c n_v^2 \log n_c)$. In this optimization problem, the number of constraints n_c is $N_l + 2N_w + 5$, while there are $n_v = N_l + 2N_w + 3$ variables. Thus, the computational complexity for solving the GP problem is

$$C_{GP} = \mathcal{O}((N_l + N_w)^3 \log(N_l + N_w)). \tag{6.5}$$

The total computational complexity of each iteration of the proposed algorithm, is the summation of the computational complexity of two steps. However, as the complexity order of the first step, i.e., converting the CGP to the GP using AGMA

---

[3] In addition to constraints explained in (6.4), $X \leq 1$ and $P \leq 1$ should also be considered to derive the solution of (6.4).

Algorithm 6 CGP-based LTE and WiFi scheduling

Input: $\Theta, \Psi, \eta, C_{\min}$
Initialization: Set initial value to $(X, Y, U, T_{\text{ra}}, v_0, x_0)$
repeat
   Step 1: Monomial approximation

     1.   Compute ζ^p for denominators of C6.4.1, C6.4.5 and C6.4.6
     2.   Use (4.29) to approximate the posynomials
     3.   Compute ζ^q for denominators of C6.4.2, C6.4.3 and C6.4.4
     4.   Use (4.30) to approximate the posynomials

   Step 2: Solve the transformed GP problem

     1.   GP-Prob $\leftarrow$ replace denominators of (6.4) with obtained monomial terms in **Step 1**
     2. $(X', Y', U', T'_{\text{ra}}, v_0', x_0') \leftarrow \text{CVX(GP-Prob)}$

until $|x_0 - x_0'| < \varsigma$
$p_{d_{\text{w}}} \leftarrow \frac{y_{d_{\text{w}}}}{\theta d_{\text{w}}(1 + y_{d_{\text{w}}})}$
Set $x_{d_{\text{l}}} = 1$ if it is the sum(X) largest elements of X, otherwise set $x_{d_{\text{l}}} = 0$
Output: X, P

(C_{APP}) is less than solving the GP problem (C_{GP}), the order of computational complexity for each iteration is equal to C_{GP}.

Note that the proposed algorithm is iterative, which solves successive GP problems until the results converge. To evaluate the required number of iterations to achieve convergence, simulations can be carried out.

6.3.3 Signaling Aspect of the Proposed Structure

The signaling needed for running Algorithm 6 by the central controller is dependent on the amount of information required to update $\theta_d(t)$ at each duty cycle t. In order to compute $\theta_d(t)$, according to Eq. (4.1), the controller needs the knowledge of the packet arrival probabilities of devices as well as the piggybacked bit information of devices. It is assumed that packet arrival probabilities of devices are fixed and can be forwarded once, thus the only information that should be received by the controller from the BS and the AP at each duty cycle is whether the devices transmitted at the previous duty cycle have more packets for transmission or not. This information can be conveyed by 1 bit for the LTE devices who were assigned a time-slot in the previous duty cycle and 2 bits for WiFi devices indicating three possible states: the device has a backlogged packet, no backlogged packet and no packet was received from the device at that duty cycle. Thus, the amount of information that should be received by the controller at each duty cycle is $N_{\text{l}} + 2N_{\text{w}}$ bits which can be considered as affordable.

6.4   Admission Control for LTE Devices

The proposed scheduling algorithm in Sect. 6.3 can satisfy a minimum requirement for the WiFi throughput at each duty cycle. However, in addition to WiFi system, the LTE devices may have quality assurance constraints. Therefore, it is important to admit a precise number of LTE devices such that their QoS requirements such as delay constraints can be met. Thus, our objective is to study the number of LTE devices that can be admitted to operate in unlicensed spectrum, while their delay requirements can be satisfied. To this end, we need to study the average packet delay for an LTE device.

6.4.1   Assumptions

To be able to study the average packet delay and derive admission control rules, we consider a homogeneous LTE network in which all devices have the same packet arrival probabilities, i.e., a_d. Thus, we omit the subscript d from a_d in the rest of the chapter. Furthermore, we assume that $\psi_d = 0$ for all devices.

To calculate the average packet delay, the pdf for the DFA length is needed. This is because the duration of the DFA segment varies at each duty cycle in the proposed scheduling algorithm, depending on the probability that devices have packets to transmit. Due to the algorithm dynamics, the derivation of pdf for DFA length is not tractable. Therefore, we focus on the *minimum* number of devices that can be supported by computing the minimum time that is assigned to the LTE system.

Algorithm 6 is run at each duty cycle t where at each duty cycle, θ is computed based on the Eq. (4.1). Let assume C_s denotes the minimum value of $C(t)$ over all t. In other words, C_s is defined as

$$C_s = \min_{t=1,2,\ldots,T} C(t), \tag{6.6}$$

if Algorithm 6 runs for $t = 1 : T$, assuming that we have a set of LTE devices with a specific set of traffic parameters, a set of WiFi devices with specific traffic parameters, a given WiFi threshold, and fixed $C_{\min}$. The following proposition describes the condition for θ which leads to $C(t) = C_s$.

Proposition 6.1 *For a saturated WiFi, having $\theta_1 = A_1$, the scheduling algorithm always assigns the airtime with duration of C_s to the LTE devices. While for the rest of duty cycles, we have $\theta_1 \geq A_1$, which leads to $C(t) \geq C_s$.*

Proof Assuming that (X', Y') is the optimal solution for $\Theta_1' = A_1$, we have

$$X'\Theta_1'^{\mathsf{T}} < X'\Theta_1''^{\mathsf{T}}, \tag{6.7}$$

where $\boldsymbol{\Theta}_1'' \geq A_1$. Following that, we can conclude that the total throughput obtains for $\boldsymbol{\Theta}_1''$ at (X', Y') is greater or equal to the total throughput of $\boldsymbol{\Theta}_1'$ at (X', Y'). Thus, for $\boldsymbol{\Theta}_1''$, the optimal solution X'' is such that we have

$$X''\mathbf{1}^\mathsf{T} \geq X'\mathbf{1}^\mathsf{T}, \tag{6.8}$$

Furthermore, $T_{ts}X''\mathbf{1}^\mathsf{T} = C''$ and $T_{ts}X'\mathbf{1}^\mathsf{T} = C_{cs}$, which leads to $C'' \geq C_s$.

Assuming a case with an unsaturated WiFi, the achievable minimum value of $C(t)$ can be obtained when LTE devices have the lowest θ_1, while WiFi devices have the highest θ_w. However, since it is not easy to derive the highest θ_w for WiFi devices, we use C_{min} for the delay analysis when we have unsaturated WiFi. The reason is that in the optimization problem (6.2), the second constraint guarantees that the duration allocated for LTE devices is larger than C_{min} time-slots, therefore it can be concluded that $C_s \geq C_{min}$.

6.4.2   Modeling the LTE Scheduling Algorithm

As mentioned above, we assume that at each duty cycle, C_{min} time-slots are dedicated to LTE devices to derive an upper bound on the delay. From the proposed scheduling algorithm, it is clear that these time-slots are assigned to devices who sent a piggybacked request in the previous duty cycle. Since these devices have packets for transmission, by assigning time-slots to them the network throughput can be maximized. However, if the number of requests is less than C_{min}, according to Proposition 6.2 the rest of time-slots are allocated to devices who have the highest θ_d among others.

Proposition 6.2 *At each duty cycle, time-slots are assigned to devices with the highest θ_d's.*

Proof The proof is similar to Proposition 5.1.

In other words, a device will obtain a time-slot either by *demand*, i.e., sending a piggyback packet (if it is currently scheduled) or as *free-assignment*, i.e., waiting for its turn to be served (if it has no assigned time-slot in the current duty cycle). Consequently, time-slot assignment to the device can be modeled by intermittently attending two servers: called server and free server. The server is designated as the called server if it is available to the device as a result of a demand assignment. Otherwise, it is called free server for the case of a free-assigned time-slot.

We now discuss the procedure that a device obtains a free time-slot. According to Eq. (4.1), θ_d depends on the a_d and $v_d(t)$. However, since we assume that devices are homogeneous and have equal a_d, $\theta_d(t)$ is only dependent on $v_d(t)$, which indicates the last time that the device has been assigned a time-slot. Larger $v_d(t)$ results in larger $\theta_d(t)$. Thus, among all devices that have not been served by the called server, the time-slot goes to the one that has not received a time-slot for the longest time in the most recent duty cycles. On the other hand, those devices that have been assigned time-slots more recently, have the lowest $\theta_d(t)$. In other words, the order that devices would be served by the free server can be modeled by a queuing system, in which the device is shifted in the end of queue whenever it is served either by the called server or the free server.

6.4.3   Delay Analysis

Since we study a homogeneous LTE system, all devices experience the same level of performance. Therefore, our derivations will be from the view point of a single device. First, we define two events that are useful to perform the delay analysis. Event E_f occurs if the forthcoming server is a free server (the next transmission of the device happens when it is her turn and there is a free time-slot), while in event of E_c the forthcoming server is a called server.

The type of the forthcoming server depends on whether the queue of the device was empty or nonempty when the precedent server departed from the device. If the device queue was empty at the departure instant of the precedent server, then the device did not request for time-slot allocation. Therefore, the forthcoming server will be a free server and used by at most one packet at the front of the queue.

Packet arrivals happen either at event E_f or E_c. Therefore, for a specific packet, the pmf of the packet delay (denoted by ϖ) can be computed as

$$\mathbb{P}(\varpi = \varpi) = \mathbb{P}(\varpi = \varpi | E_f)\mathbb{P}(E_f) + \mathbb{P}(\varpi = \varpi | E_c)\mathbb{P}(E_c), \tag{6.9}$$

where $\mathbb{P}(\varpi = \varpi | E_f)$ and $\mathbb{P}(\varpi = \varpi | E_c)$ are conditional probabilities of delay given E_f and E_c [13].

To calculate $\mathbb{P}(\varpi = \varpi)$ based on (6.9), in the following, we first explain how to obtain p_0 (the probability of empty device queue), which facilitates computing $\mathbb{P}(\varpi = \varpi | E_f)$ and $\mathbb{P}(\varpi = \varpi | E_c)$, and finally we derive $\mathbb{P}(E_f)$ and $\mathbb{P}(E_c)$.

6.4.3.1   Derivation of p_0

In order to derive p_0, we define r'_k that represents the number of remaining packets in the queue at the departure instant of k (i.e., the time that the device leaves the server).

First, let assume that the device will be served by a free server. This happens when the device queue was empty at the departure instant of the precedent server (i.e., $r'_{k-1} = 0$). We define the random variable z as the number of packet arrivals between the departure instants of $k - 1$ and k. Therefore, in this case, $r'_k = \max\{z - 1, 0\}$. Subsequently, r'_k will be zero if at most one packet arrives between the departure instants of $k - 1$ and k. This is because only one packet can be served by the free server. Thus, $\mathbb{P}(r'_k = 0 | E_f)$ can be computed as

$$\mathbb{P}(r'_k = 0 | E_f) = \sum_{l=l_{\min}}^{+\infty} [\mathbb{P}(z=0) + \mathbb{P}(z=1)] L_f(l, N_1) \qquad (6.10)$$

where $L_f(l, N_1)$ represents $\mathbb{P}(\mathbf{l} = l | \mathbf{n} = N_1)$, i.e., the probability that it takes l time-slots for the device to be served by a free server after the last departure from the precedent server, given that N_1 devices exist in the system. Moreover, $l_{\min}$ represents the feasible lower bound of l that can be computed as

$$l_{\min} = [\frac{N_1}{N_{C_{\min}}}] N_{\mathrm{dc}} + N_1 \bmod N_{C_{\min}} \qquad (6.11)$$

where $x \bmod y$ represents the remainder of x divided by y, $N_{\mathrm{dc}} = [T_{\mathrm{dc}}/T_{\mathrm{ts}}]$ and $N_{C_{\min}} = C_{\min}/T_{\mathrm{ts}}$. This value is realized if each device has at most one packet for transmission at the time it is allocated a time-slot. We can derive $\mathbb{P}(z = 0)$ and $\mathbb{P}(z = 1)$ as

$$\mathbb{P}(z=0) = \sum_{l=l_{\min}}^{\infty} (1-a)^{\frac{l}{N_{\mathrm{dc}}}} L_f(l, N_1) \qquad (6.12)$$

$$\mathbb{P}(z=1) = \sum_{l=l_{\min}}^{\infty} \frac{l}{N_{\mathrm{dc}}} a (1-a)^{\frac{l}{N_{\mathrm{dc}}}-1} L_f(l, N_1) \qquad (6.13)$$

Thus, $\mathbb{P}(r'_k = 0 | E_f)$ can be obtained as

$$\mathbb{P}(r'_k = 0, E_f) = \sum_{l=l_{\min}}^{\infty} \left[(1-a)^{\frac{l}{N_{\mathrm{dc}}}} + \frac{l}{N_{\mathrm{dc}}} a(1-a)^{\frac{l}{N_{\mathrm{dc}}}-1} \right] L_f(l, N_1)$$

Now, assume that the next server is a called server. Then, $\mathbb{P}(r'_k = 0, E_c)$ can be computed as

$$\mathbb{P}(r'_k = 0 | E_c) = \sum_{l=l_{\min}}^{+\infty} \sum_{z=2}^{+\infty} \sum_{i=1}^{l-N_{\mathrm{dc}}} \mathbb{P}(z=z) L_f(l, N_1) \qquad (6.14)$$

where

$$\mathbb{P}(z=z) = \frac{l-i}{l N_{\mathrm{dc}}} \binom{[\frac{l-i}{N_{\mathrm{dc}}}] + z - 2}{k-1} a^z (1-a)^{\frac{l}{N_{\mathrm{dc}}}} \qquad (6.15)$$

In order to derive the probability that the device leaves the called server (i.e., $\mathbb{P}(r'_k = 0 | E_c)$), we take a summation over the all possibilities of z. For $z = z$, the device leaves the called server after z duty cycles from being served by the free server. In other words, it means that z packets arrived at the device's queue during the time that the device was waiting for the free server and the time it spent in the called server. Furthermore, the last packet should arrive at the last duty cycle before the departure, otherwise the device will depart from the server after $z - 1$ duty cycles.

Finally, p_0 can be found as

$$p_0 = \mathbb{P}(r'_k = 0|E_f)\mathbb{P}(E_f) + \mathbb{P}(r'_k = 0|E_c)\mathbb{P}(E_c) \tag{6.16}$$

6.4.3.2   Derivation of $L_f(l, n)$

In the proposed scheme, after a device departs the free server, it will be served by the free server again if during this interval, all the other devices have left either free server or called server. With the assumption that at each duty cycle $N_{C_{\min}}$ devices are served, those devices that are currently allocated a time-slot will be served in the subsequent duty cycles until their queues become empty. Thus, $L_f(l, n)$ for $l > N_{dc}$ can be calculated as

$$L_f(l, n) = \sum_{i=0}^{M} \binom{M}{i} {p_0}^i (1 - p_0)^{M-i} L_f(l - N_{dc}, N_1 - i) \tag{6.17}$$

where $M = \min(n, N_{C_{\min}})$. Furthermore, for $l < N_{dc}$ we have

$$L_f(l, n) = \begin{cases} p_0^n & \text{if } l = n \\ 0 & \text{otherwise} \end{cases} \tag{6.18}$$

The recursive equation (6.17) indicates that at each duty cycle, i from M devices leave the server. Thus, in order to have $1 = l$, the remaining number of devices (which is $N_1 - i$) should leave the server in $l - N_{dc}$ time-slots. It is obvious that for $l < N_{dc}$, only if the number of time-slots is equal to the number of devices with probability of p_0^n, all devices depart the server. Otherwise, i.e., $l \neq n$, it means either devices left the server earlier than l or later. In both cases, the probability is equal to zero.

Now that L_f is derived, we can calculate p_0. To this end, we use Eq. (6.16) in which all terms are expressed in terms of p_0. To solve this one variable equation, numerical methods can be applied.

6.4.3.3   Derivation of $\mathbb{P}(E_f)$ and $\mathbb{P}(E_c)$

To derive $\mathbb{P}(E_f)$ and $\mathbb{P}(E_c)$, we need to compute $\bar{M}_f$ and $\bar{M}_c$ which represent the average length of events E_f and E_c, respectively.

$$\bar{M}_f = \sum_{l=l_{min}}^{\infty} l L_f(l, N_1) \tag{6.19}$$

and

$$\bar{M}_c = \sum_{z=2}^{\infty} \sum_{l=l_{min}}^{\infty} \sum_{i=1}^{l-N_{dc}} z \frac{l-i}{l} \binom{\lceil \frac{l-i}{N_{dc}} \rceil + z - 2}{k-1} a^z (1-a)^{\frac{l}{N_{dc}}} L_f(l, N_1) \tag{6.20}$$

Based on [14], $\mathbb{P}(E_f)$ and $\mathbb{P}(E_c)$ can be derived as

$$\mathbb{P}(E_f) = \frac{p_0 \bar{M}_f}{p_0 \bar{M}_f + (1-p_0)\bar{M}_c} \qquad \mathbb{P}(E_c) = \frac{(1-p_0)\bar{M}_c}{p_0 \bar{M}_f + (1-p_0)\bar{M}_c} \tag{6.21}$$

6.4.3.4   Average Packet Delay

Here, we first derive the conditional probability of delay given E_f (i.e., $\mathbb{P}(\varpi = \varpi | E_f)$). Assuming that the length of an interval between two departures is l and the arrival time of the first packet is at i, the packet delay can be computed for two cases as follows

- **First packet case**: The delay is equal to $(l - i)$ if the packet is the first one to arrive in the queue in the interval. We call this event F_1.
- **Non-first packet case**: In this case, denoted by F_2, there are some packets already in the queue. Therefore, the packet will be transmitted, if all packets ahead of it in the queue are scheduled first. In this situation, the first packet of queue is served by the free server. Since the queue is not empty, the piggyback bit will be set to 1. Thus, the remaining packets will be served by the called server. Consequently, the packet delay will be the summation of two parts: $(l - i)$ that is the waiting time required for the first packet to be served by the free server and the second part is the waiting time needed for serving the rest of packets by the called server.

Based on these two cases, the probability of packet delay can be computed as

$$\mathbb{P}\left(\varpi = \varpi | E_f\right) = \sum_{l=l_{min}}^{+\infty} \mathbb{P}\left(\varpi = \varpi | l = l, E_f\right) L_f(l, N_1) \tag{6.22}$$

where

$$\mathbb{P}\left(\varpi = \varpi | l = l, E_f\right) = \mathbb{P}\left(\varpi = \varpi, F_1 | l = l, E_f\right) + \mathbb{P}\left(\varpi = \varpi, F_2 | l = l, E_f\right)$$

Subsequently, we derive

$$\mathbb{P}\left(\boldsymbol{\varpi} = \varpi, F_1 | \mathbf{l} = l, E_f\right) = \frac{1}{l}\mathbb{P}(\text{no arrival in } l - \varpi \text{ time-slots}) \qquad (6.23)$$

and

$$\mathbb{P}\left(\boldsymbol{\varpi} = \varpi, F_2 | \mathbf{l} = l, E_f\right)$$

$$= \sum_{i=1}^{l} \sum_{h=2}^{\frac{[l-i]}{N_{dc}}} \frac{1}{l}\mathbb{P}\left(h \text{ arrivals in } [l - i - \varpi + hN_{dc}] \text{ time-slots}\right)$$

$$\mathbb{P}(\text{no arrival in } l - \varpi \text{ time-slots})$$

$$= \sum_{i=1}^{l} \sum_{h=2}^{\frac{[l-i]}{N_{dc}}} \frac{1}{l}\binom{\frac{l-i-\varpi}{N_{dc}}+h}{h} a^h (1 - a)^{(l-\varpi)/N_{dc}}. \qquad (6.24)$$

The packet can also arrive at Event E_c, in this case the delay can be computed as

$$\mathbb{P}\left(\boldsymbol{\varpi} = \varpi | \mathbf{l} = l, E_c\right) = \mathbb{P}\left(r_k' = \varpi/N_{dc}\right)$$

$$= \sum_{i=1}^{l} \frac{1}{l}\mathbb{P}\left(\varpi/N_{dc} \text{ arrivals in } [l - i + \varpi - 2N_{dc}] \text{ time-slots}\right)\mathbb{P}(\text{no arrival in } l \text{ time-slots})$$

$$= \sum_{i=1}^{l} \frac{l-i}{lN_{dc}}\binom{\frac{l-i}{N_{dc}}+\varpi/N_{dc}-2}{\varpi/N_{dc}-1} a^{\varpi/N_{dc}} (1 - a)^{l/N_{dc}}, \qquad (6.25)$$

which means that in order to have a delay equals to ϖ, the packet must be arrived when there are ϖ/N_{dc} packets at the device queue.

6.5 Illustrative Results

For performance evaluation, we consider a system with one LTE BS and one WiFi AP, both operating on the same channel. We obtain the results in Matlab environment and we use CVX to derive the solution of GP problems [15]. We assume that $T_{dc} = 100\,\text{ms}$, each time-slot is equal to $T_{ts} = 6\,\text{ms}$ and the time unit duration is $10\,\mu\text{s}$. In the following, we present the considered scenarios for performance evaluation along with their results.

6.5.1 Effect of Increasing N_l

We first investigate how increasing the number of devices in the LTE system can affect the WiFi throughput and the overall network throughput. For this scenario, we assume that the WiFi throughput threshold η is equal to 4 time-slots. We consider a case where $N_w = 14$ and packet arrival probabilities are $A_l = \{[0.8]_4, [0.5]_{N_l-4}\}$

and $A_\mathrm{w} = \{[0.8]_4, [0.5]_{10}\}$. More specifically, this means that the WiFi network serves 4 devices with $a_d = 0.8$ and 10 devices with $a_d = 0.5$. Similarly, LTE has 4 devices with $a_d = 0.8$ and the rest of its devices are the ones with $a_d = 0.5$. Furthermore, we assume that devices of both networks are randomly located in a circular region with radios of 5 m, and channel parameters are as following: path loss exponent $\zeta = 3$, receiver threshold $\xi_R = 0\,\mathrm{dB}$ and $\frac{p_t}{\sigma_n^2} = 20\,\mathrm{dB}$.

We compare the performance of the proposed algorithm with the Fixed C algorithm as a benchmark to verify the effectiveness of our approach. In the Fixed C algorithm, at each duty cycle a fixed number of time-slots is assigned to the LTE devices with the highest expected throughput, i.e., $\theta(1 - \psi)$. Furthermore, in this algorithm, WiFi devices compete with each other using fixed p parameters. Here, we assume that $C = 5$ time-slots, and p is 0.05 for all WiFi devices.

As can be seen in Fig. 6.2, by increasing the number of LTE devices, i.e., N_1, greater throughput can be achieved for both LTE system and the overall network. The reason is that more packets are generated, therefore with higher probability time-slots are allocated to the devices who have packets for transmission. On the other hand, WiFi throughput decreases because greater throughput can be achieved by assigning more time-slots to the LTE system. However, due to the WiFi throughput constraint, WiFi throughput never falls below the targeted threshold. On the other hand, for the Fixed C algorithm, the LTE and overall throughput remains the same, the reason is that number of time-slots assigned to these devices is fixed

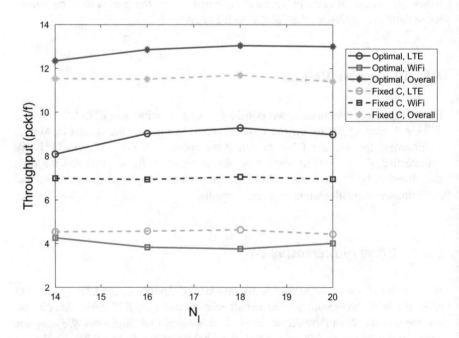

Fig. 6.2 Throughput versus N_1

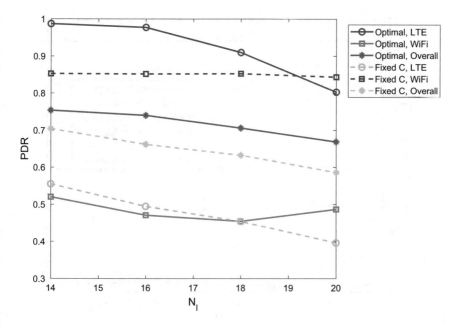

Fig. 6.3 PDR versus N_l

and since these time-slots are allocated to the LTE devices with the highest expected throughput, LTE throughput does not change by increasing N_l.

Furthermore, the effect of increasing N_l on the packet delivery ratio (PDR) are demonstrated in Fig. 6.3. PRD is defined as the ratio of number of transmitted packets to the number of generated packets. As observed, the proposed algorithm outperforms the Fixed C algorithm in terms of the LTE and overall PDR. Furthermore, for $N_l = \{14, 16\}$, LTE PDR is so close to 1, however by increasing N_l it starts to drop, which means that LTE devices' quality of service requirements would be affected. In order to avoid this situation, the number of LTE devices should be controlled otherwise LTE devices will suffer from a performance degradation.

6.5.2   Effect of Increasing N_w

In another scenario, we have the results for increasing the number of devices in the WiFi network, where $A_l = \{[0.8]_4, [0.5]_{10}\}$ and $A_w = \{[0.8]_4, [0.5]_{N_w-4}\}$ and the rest of the parameters are the same as Sect. 6.5.1. In Fig. 6.4, it is evident that in both LTE and WiFi, throughput remains unchanged for the optimal algorithm. In fact, when LTE system has high-traffic devices, higher throughput can be achieved by allocating more time to them, since for high-traffic devices, the DFA scheme achieves better performance compared to the CSMA. The reason is that

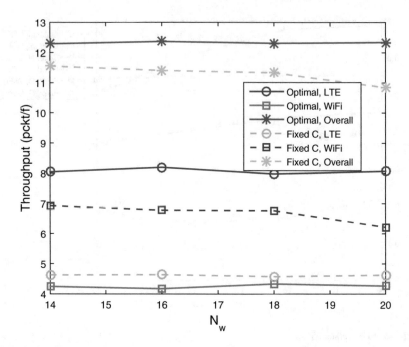

Fig. 6.4 Throughput versus N_w

in CSMA, due to the backoffs and collisions, time would be wasted. Therefore, in Fig. 6.4 where LTE devices have high traffic, the following statement is true. The higher throughput could be achieved if more time slots were allocated to LTE. Consequently, the algorithm is reluctant to add more time slots for WiFi system since its throughput requirement is already satisfied, even if the number of WiFi devices is increasing. The other point is that increasing N_w may lead to a larger number of collisions, therefore, to meet the WiFi throughput threshold, lower p probabilities are assigned by the optimal algorithm to the devices. However, the WiFi throughput of the Fixed C algorithm drops by increasing N_w, as p values are fixed.

Furthermore, as shown in Fig. 6.5 the WiFi PDR of the optimal algorithm decreases by growing number of WiFi devices. The reason is that increasing N_w leads to the larger number of packets, while the number of transmitted packets remains the same.

6.5.3   Homogeneous LTE Network

Here, we obtain the results for homogeneous LTE network. We derive the results for $C_{\min} = 5$ and 7 time-slots. Moreover, we consider the packet arrival probabilities

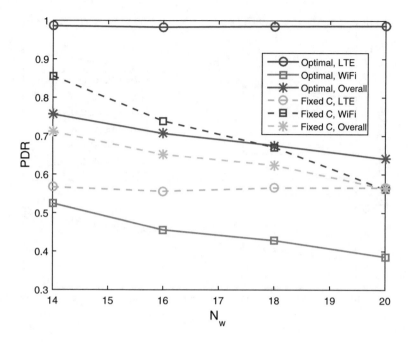

Fig. 6.5 PDR versus N_w

as $A_l = \{[0.3]_{N_l}\}$ and $A_w = \{[1]_8\}$. As observed from Fig. 6.6, by increasing N_l, LTE throughput increases as well. The reason is that since more number of packets are generated, less time-slots are left idle. Moreover, although more number of time-slots are assigned to the LTE network for $C_{min} = 7$, but same throughput is achieved for two values of C_{min}. The reason is that LTE network is underutilized, therefore adding more time-slots does not lead to greater throughput. However, for $C_{min} = 5$ more time-slots are left for WiFi and since WiFi devices are saturated, the increased time-share leads to larger throughput.

Furthermore, we obtain delay results for this scenario shown in Fig. 6.7. The results are compared with the analytical upper-bound which is derived for saturated WiFi by assigning C_{min} time-slots to LTE devices at each duty cycle. As observed, the gap between the upper-bound and the proposed scheduling algorithm increases as N_l grows. The reason is that by increasing N_l, more traffic is generated in the LTE network. Therefore, greater throughput can be achieved by assigning more time-slots to LTE devices. Thus, the probability that $C(t) > C_{min}$ increases which makes the gap between the proposed scheme and the upper-bound becomes larger.

Furthermore, for $C_{min} = 7$ the upper bound delay and simulation results are close, while for $C_{min} = 5$, the gap is larger. The reason is that with $C_{min} = 7$, in most of the duty cycles, the derived $C(t)$ is equal to C_{min}. However, $C_{min} = 5$ leads to $C(t) > C_{min}$ more frequently.

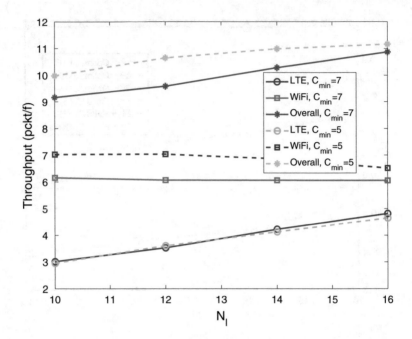

Fig. 6.6 Throughput versus N_l

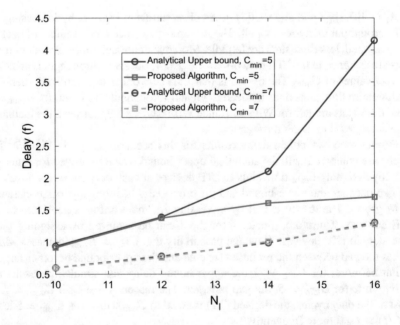

Fig. 6.7 Delay versus N_l

Moreover, as observed, $C_{min} = 7$ achieves smaller delay compared to the $C_{min} = 5$. Since larger C_{min} imposes to assign more time-slots to the LTE network. Therefore, at each duty cycle, more number of packets are served which leads to shorter delays for LTE devices. However, for case of $N_l = 16$, the same amount of delay is achieved for both values of C_{min}. Since for this case, LTE network is more loaded, therefore in most of the duty cycles, we have $C(t) > C_{min}$, meaning that the output of the scheduling is not dependent on C_{min}.

6.5.4 Computational Complexity

Here, we investigate the average number of iterations required for Algorithm 6 to converge. The results in Fig. 6.8 are obtained for $A_l = A_w = \{[0.8]_4, [0.5]_{N_d/2-4}\}$, where $N_d = N_l + N_w$. Furthermore, for this scenario, we set $\eta = 2$ time-slots, the convergence parameter $\varsigma = 0.01$ and the same setting for channel parameters as Sect. 6.5.1. As observed, the average number of iterations grows only linearly with N_d. It should be noted that the average number of iterations also depends on the convergence parameter. Lower average number of iterations can be achieved for larger ς.

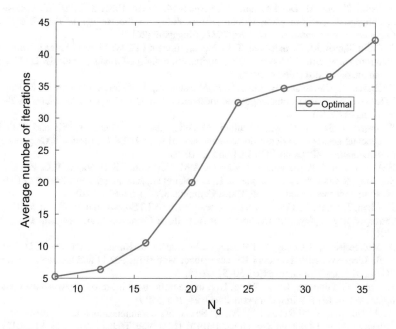

Fig. 6.8 Average number of iterations versus N_d

6.6   Concluding Remarks

In order to satisfy the increasing demand for mobile traffic, LTE operation over unlicensed bands has been proposed. In this chapter, we have considered the scenario that both LTE and WiFi systems share the same unlicensed band. In such a setting, the main challenge for LTE deployment is that the performance of WiFi system should not degrade significantly. In order to address this issue, we consider a coordinated approach in which both systems are connected to a central network entity. This entity manages the channel access between these two systems such that the overall spectrum efficiency is improved while the WiFi performance does not fall below a certain level. In order to reach this goal, a duty-cycle-based approach is used, in which the time is divided into duty cycles and the exclusive share of each system is dynamically optimized by the network entity. It is shown that the developed algorithm can ensure a minimum throughput requirement for WiFi while maximizing the total throughput. Furthermore, we obtain an upper-bound for average delay of LTE devices. Using this analysis, we can derive the minimum number of LTE devices that can be admitted by the network while their delay requirements are met.

References

1. A. Dalili Shoaei, M. Derakhshani, T. Le-Ngoc, M. Salem, Efficient LTE/WiFi coexistence in unlicensed spectrum using virtual network entity, in *Proceedings of the IEEE Global Communication Conference (GLOBECOM)*, Singapore (2017)
2. A. Dalili Shoaei, M. Derakhshani, T. Le-Ngoc, Efficient LTE/Wi-Fi coexistence in unlicensed spectrum using virtual network entity: optimization and performance analysis. IEEE Trans. Commun. **66**(6), 2617–2629 (2018)
3. F. Chaves, A. Cavalcante, E. Almeida, F. Abinader Jr., R. Vieira, S. Choudhury, K. Doppler, LTE/Wi-Fi coexistence: challenges and mechanisms, in *XXXI Simposio Brasileiro de Telecomunicacoes* (2013)
4. S. Sagari, I. Seskar, D. Raychaudhuri, Modeling the coexistence of LTE and WiFi heterogeneous networks in dense deployment scenarios, in *IEEE International Conference on Communication Workshop (ICCW)*, London (2015)
5. F.M. Abinader, E.P. Almeida, F.S. Chaves, A.M. Cavalcante, R.D. Vieira, R.C. Paiva, A.M. Sobrinho, S. Choudhury, E. Tuomaala, K. Doppler et al., Enabling the coexistence of LTE and Wi-Fi in unlicensed bands,. IEEE Trans. Commun. **52**(11), 54–61 (2014)
6. B. Chen, J. Chen, Y. Gao, J. Zhang, Coexistence of LTE-LAA and Wi-Fi on 5 GHz with corresponding deployment scenarios: a survey. IEEE Commun. Surv. Tutorials **19**(1), 7–32 (2017)
7. A. Mukherjee, J.-F. Cheng, S. Falahati, H. Koorapaty, R. Karaki, L. Falconetti, D. Larsson et al., Licensed-assisted access LTE: coexistence with IEEE 802.11 and the evolution toward 5G. IEEE Trans. Commun. **54**(6), 50–57 (2016)
8. H. Cui, V.C. Leung, S. Li, X. Wang, LTE in the unlicensed band: overview, challenges, and opportunities. IEEE Wirel. Commun. **24**(4), 99–105 (2017)
9. A. Al-Dulaimi, S. Al-Rubaye, Q. Ni, E. Sousa, 5G communications race: pursuit of more capacity triggers LTE in unlicensed band. IEEE Trans. Veh. Technol. **10**(1), 43–51 (2015)

10. Q. Chen, G. Yu, H.M. Elmaghraby, J. Hamalainen, Z. Ding, Embedding LTE-U within Wi-Fi bands for spectrum efficiency improvement. IEEE Netw. **31**(2), 72–79 (2017)

11. G. Xu, Global optimization of signomial geometric programming problems. Eur. J. Oper. Res. **233**(3), 500–510 (2014).

12. M. Derakhshani, X. Wang, D. Tweed, T. Le-Ngoc, A. Leon-Garcia, AP-STA association control for throughput maximization in virtualized WiFi networks. IEEE Access **6**, 45034–45050 (2018)

13. T. Le-Ngoc, I.M. Jahangir, Performance analysis of CFDAMA-PB protocol for packet satellite communications. IEEE Trans. Commun. **46**(9), 1206–1214 (1998)

14. D.R. Cox, *Renewal Theory*, vol. 58 (Methuen, 1962)

15. M. Grant, S. Boyd, CVX: Matlab software for disciplined convex programming, version 2.1 (2014). http://cvxr.com/cvx

Chapter 7
A NOMA-Enhanced Reconfigurable Access Scheme with Device Pairing for MTC

7.1 Introduction

In this chapter, we focus on the massive MTC scenario with short packet sizes, which pose major challenges on the network optimization and multiple access [1, 2]. To date, some potential candidates have been proposed to support massive connectivity, such as massive multiple-input-multiple-output (MIMO), millimeter wave communications, ultra dense networks, and non-orthogonal multiple access (NOMA). In this chapter, we adopt NOMA, which is highly expected to increase the network throughput and accommodate massive connectivity.

NOMA allows multiple devices to transmit over the same resource simultaneously using power-domain or code-domain techniques, while in conventional OMA schemes, radio resources are orthogonally assigned to devices to avoid or alleviate interference [3]. Accordingly, it is expected that by using NOMA, the network throughput significantly increases compared to using OMA schemes, since if each resource is simultaneously used by multiple devices, the total network efficiency becomes multiple folds too. However, in M2M networks, since devices might have sporadic transmissions, even though a resource is shared among them, it may still left unused, leading to low spectrum utilization. In other words, the NOMA scheme is more useful for devices having periodic or frequent packets for transmissions compared to sporadic transmissions for which random access schemes are more pertinent.

In this chapter, to achieve the spectral efficiency optimality, we propose a NOMA-enhanced reconfigurable access scheme (NERA) which is able to switch between the assignment-based NOMA-OMA and random access schemes, taking into account the device traffic statistics.

To obtain the optimal length of each segment, we formulate a network throughput maximization problem such that the device pairing in the NOMA segment can be optimized under minimum rate requirements of the devices. Due to the combina-

© Springer Nature Switzerland AG 2020
T. Le-Ngoc, A. Dalili Shoaei, *Learning-Based Reconfigurable Multiple
Access Schemes for Virtualized MTC Networks*, Wireless Networks,
https://doi.org/10.1007/978-3-030-60382-3_7

torial nature of the formulated mixed integer non-linear programming (MINLP) problem, we propose a decomposition-based scheme which solves the problem into two steps. In the first step, to choose a set of paired devices for the NOMA segment, we formulate the problem as a weighted matching problem and propose a sub-optimal algorithm to solve it. Then, in the next step, for a given set, we obtain the length of each segments such that the total network throughput is maximized. The second step is formulated as an optimization problem which is solved in an iterative manner. The performance of the proposed scheme is compared with a reconfigurable scheme which does not support NOMA. Considering different scenarios, the results show when using the NOMA-enhanced scheme is beneficial.

The rest of this chapter is organized as follows. Section 7.2 presents the system model under consideration along with the frame structure of the proposed NOMA-enhanced reconfigurable scheme. In Sect. 7.3, an optimization problem is formulated to maximize the throughput of the proposed scheme. Due to the computational complexity of the problem, it is divided into two sub-problems, and the solution of the sub-problems are provided in Sects. 7.4 and 7.5. Section 7.6 presents the simulation results to demonstrate the efficacy of the proposed scheme. Finally, Sect. 7.7 concludes the chapter.

7.2   System Model

We consider an M2M network consisting of one AP and N_d devices. We assume that all devices transmit their packets with the same transmission power p_t. The channel power gain coefficient between device $d \in \mathcal{D} = \{1, \cdots, N_d\}$ and the AP is g_d and the received power from device d at the AP is $p_t g_d + n_d$, where n_d denotes the Gaussian noise power. The traffic model of devices is as described in Chap. 4, Sect. 4.2.2, where a_d denotes the probability that a new packet is added to the queue of device d. We assume that the AP is aware of these packet arrival probabilities and at each frame it updates θ_d, for each $d \in \mathcal{D}$, which is the probability that device d has a packet for transmission, according to Eq. (4.1).

7.2.1   NOMA with Imperfect SIC

In Chap. 2, Sect. 2.3.1, we have provided an overview of the uplink NOMA and we have explained how SIC technique can be used in order to detect received signals at the AP. Here, we explain the situation that perfect SIC cannot be guaranteed.

In the case of the imperfect SIC, when the device's signal is decoded, there is a difference between the actual and estimated signal. In other words, in the imperfect SIC, some portion of the received power of devices remains as interference which is called the residual interfering signal power and it is denoted by I_d^{ri}. The magnitude of the SIC error is dependent on the type of SIC employed, the number of signals being canceled, and channel and device mobility conditions [4]. Thus, unlike the perfect SIC, where the signal of the lowest channel gain device is decoded with no interference from other devices, here it contains the residual interfering signal powers from devices with larger channel gains. To consider all sources of error, we define the expected level of cancellation achieved by SIC as $\sigma^2 = \mathbb{E}[|s_d - \hat{s}_d|]$, where s_d and $\hat{s}_d$, are the received and estimated signals of device d at the AP, respectively. Thus, we have

$$I_d^{\mathrm{ri}} = \sigma^2 p_t g_d. \tag{7.1}$$

7.2.2 NOMA-Enhanced Reconfigurable Access Scheme

The network operates on a frame-by-frame basis. The proposed frame structure is similar to the one considered in Chap. 4, however, to support NOMA transmissions, another segment has been added to that. In particular, each frame is started with a beacon followed by three traffic segments: NOMA-based demand-free assignment (NDFA), OMA-based demand-free assignment (ODFA), and random access (RA) as shown in Fig. 7.1. The beacon is transmitted to the devices by the AP for notifying them on the scheduling of the frame. The three traffic segments have a total of N_{ts} time-slots and each time-slot has a duration of T_{ts}. Thus, the length of each frame denoted by T_f is $N_{\mathrm{ts}}T_{\mathrm{ts}}$. The NDFA and ODFA have a total length (time) of less than T_{max}. Operations in the three traffic segments are described in the following paragraphs.

Beacon At the start of each time frame, the AP broadcasts a beacon frame including the information about the assignment-based NOMA-OMA and random access schemes.

NOMA-Based Demand-Free Assignment (NFDA) In this segment, the granted devices transmit their packets in the corresponding allocated time-slots using NOMA. Here, we consider the case that only up to two devices can simultaneously transmit their packets in an allocated time-slot. In fact, as the received signal at the AP is superposition of at most two signals, the decoding complexity is acceptable.

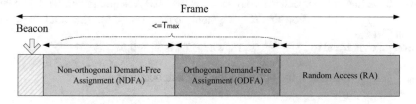

Fig. 7.1 Frame structure for the proposed NOMA-enhanced reconfigurable access scheme

OMA-Based Demand-Free Assignment (ODFA) In this traffic segment, similar to the NDFA, granted devices transmit their packets in the corresponding allocated time-slots. The only difference is that, each time-slot is only assigned to one device.

Random Access (RA) In this traffic segment, devices that have packets for transmission but are not granted a time-slot, contend with each other, based on the p-persistent CSMA protocol, where p_d parameter of each device to use this scheme is announced by the AP in the beacon.

7.3   Problem Formulation

NOMA can provide higher throughput for MTC, as it can serve multiple devices using the same radio resource. However, in MTC, devices may transmit in a sporadic manner, where random access schemes may have better performance. This is because the allocated resources might be left idle due to their low rate of transmissions. Thus, deploying a combination of NOMA and random access scheme in each frame can be beneficial, if devices with a higher probability of transmission are scheduled to transmit in the assignment-based segments and the rest for the random access segment. In fact, using this scheme, the network throughput depends on which devices are selected for the NOMA segment and how these devices are paired. In the following, we present the problem formulation for the scheduling of the proposed scheme.

7.3.1   Optimization Problem

To schedule devices based on the proposed reconfigurable access scheme, we formulate an optimization problem aiming to maximize the network throughput. Define $x_{d,d'}^{n}$ as a binary variable indicating whether devices d and d' are paired together to be assigned a time-slot in the NDFA segment. In NOMA, devices are said to be in a pair, if they simultaneously use the same time-slot. In fact, the efficiency of the NOMA scheme depends on how devices are paired. With a constraint that the rate of device d is sufficiently large to transmit one packet during one time-slot, the achievable expected throughput for the NDFA segment can be calculated as $\sum_{d \in \mathcal{D}} \sum_{d' \in \mathcal{D}} \theta_d x_{d,d'}^{n}$. For the RA segment, based on (4.8), the total expected throughput is equal to $\sum_{d \in \mathcal{D}} \sum_{d' \in \mathcal{D}} \rho_d T_{\text{ra}}$ where T_{ra} is $T_f - 0.5 T_{\text{ts}} \sum_{d \in \mathcal{D}} \sum_{d' \in \mathcal{D}, d \neq d'} x_{d,d'}^{n} + T_{\text{ts}} \sum_{d \in \mathcal{D}} x_{d,d}^{n}$. Therefore, the optimization problem becomes

$$\max_{X^{n},Y} \sum_{d\in\mathcal{D}} \left[\sum_{d'\in\mathcal{D}} \theta_d x^{n}_{d,d'} + \rho_d T_{\mathrm{ra}} \right] \tag{7.2}$$

subject to:

C7.2.1: $\log_2 \left(1 + \dfrac{p_t\gamma_d}{p_t\gamma_{d'} + n_d} \right) \geq Rx^{n}_{d,d'}, \ \forall d \in \mathcal{D}, \ \forall d' < d,$

C7.2.2: $\log_2 \left(1 + \dfrac{p_t\gamma_d}{\sigma^2 p_t\gamma_{d'} + n_d} \right) \geq Rx^{n}_{d,d'}, \ \forall d \in \mathcal{D}, \ \forall d' > d,$

C7.2.3: $\log_2 \left(1 + \dfrac{p_t\gamma_d}{n_d} \right) \geq Rx^{n}_{d,d}, \ \forall d \in \mathcal{D},$

C7.2.4: $y_d \displaystyle\sum_{d'\in\mathcal{D}} x^{n}_{d,d'} = 0, \qquad \forall d \in \mathcal{D},$

C7.2.5: $0.5T_{\mathrm{ts}} \displaystyle\sum_{d\in\mathcal{D}} \sum_{d'\in\mathcal{D}} x^{n}_{d,d'} + T_{\mathrm{ts}} \sum_{d\in\mathcal{D}} x^{n}_{d,d} \leq T_{\max},$

C7.2.6: $\dfrac{y_d}{\theta_d(1 + y_d)} \leq 1, \qquad \forall d \in \mathcal{D},$

C7.2.7: $x^{n}_{d,d'} \in \{0, 1\}, \qquad \forall d \in \mathcal{D}, \ \forall d' \in \mathcal{D}.$

In C7.2.1, R is the required rate for transmitting a packet in one time-slot. In the SIC, the AP first decodes the signal from the device with the higher channel gain. Consequently, the higher channel gain device experiences interference from its paired device. Assuming that devices are sorted in ascending order with respect to their channel gains, constraint C7.2.1 ensures that the rate of device d with higher channel gain is sufficiently large to transmit one packet during one time-slot. Constraint C7.2.2 guarantees the minimum rate requirement of the paired device with the weaker signal, which in the case of imperfect SIC suffers from residual interfering signal power. C7.2.3 ensures that the required rate of the ODFA device is met. Constraint C7.2.4 indicates that device d is either selected for NDFA, ODFA or RA segment. Constraint C7.2.5 guarantees that the duration of assignment-based access is less than $T_{\max}$. Constraint C7.2.6 indicates that the p_d of each device for the CSMA scheme should be less than 1. Finally, constraint C7.2.7 defines $x^{n}_{d,d'}$ as a binary variable.

The problem in (7.2) is a mixed-integer optimization problem due to the binary constraint of C7.2.7. Furthermore, it is non-convex, as the objective function and constraint C7.2.7 are non-convex. Consequently, it is a non-convex mixed-integer optimization problem. Generally, there is no computational efficient approach to solve this class of optimization problems. However, the problem can be decomposed into two sub-problems; (1) device pairing sub-problem, (2) NERA optimization sub-problem. The device pairing sub-problem deals with how to partition devices into the groups while the solution of NERA optimization sub-problem determines the optimal number of NOMA groups and p_d for devices in the RA segment. In the following, we describe how to obtain an efficient and tractable solution for each of these sub-problems.

7.4   Device Pairing for NOMA

In this section, we discuss the proposed device pairing scheme. In NOMA, to be able to perform SIC, devices should be paired such that the difference between their received powers at the AP is sufficiently large. Otherwise, if the difference in the received powers is small, the signals will not be successfully decoded at the AP. The device pairing sub-problem is inherently combinatorial, which requires an exhaustive search to obtain the optimal solution. Here, we propose a simple yet high performance algorithm to solve this sub-problem. More specifically, we model it as a weighted graph matching problem, which is shortly explained in the following subsection.

7.4.1   Introduction to Weighted Matching Problem

Let consider a bipartite graph $G = (V, E)$, where V and E denote the set of vertices and edges respectively and w is a function which assigns a weight to the edges of G, i.e., $w : E \longrightarrow \mathcal{R}$. A matching M is a sub-graph of G with edges of $F \subseteq E$, where no two edges in F share a common vertex. The weight of this matching is the sum of the weights in M, i.e., $w(M) = \sum_{e \in M} w(e)$. The maximum weighted matching is the matching of a graph G with the largest value of $w(M)$. In [5], Edmonds proposed an exact algorithm with running time of $O(|V|^4)$, which is not affordable computationally. The fastest algorithm for solving this problem has a running time of $O(|V||E| + |V|^2 \log |V|)$ [6], which is still costly for large graphs. Therefore, approximation algorithms have been proposed which are faster than the exact algorithms. Furthermore, the other drawback of Edmonds' algorithm is that it is non-intuitive and consequently difficult to understand and implement [7]. In the following, we discuss an approach which is simple but yet efficient.

The device pairing sub-problem can be modeled as a weighted matching problem in which the vertices denote devices and an edge between two vertices d and d' indicates that the simultaneous transmission of those devices over the same time-slot does not violate the SIC constraint, i.e., the signals of devices d and d' can be successfully decoded at the AP. Finally, the weight of each edge is the sum of the throughput of its endpoints, i.e., for edge e connecting devices d and d', the weight is

$$w(e_{d,d'}) = \theta_d + \theta_{d'}.$$

Algorithm 7 Device pairing for NOMA

  Input: $G(V, E)$, N_p
  $K \leftarrow [0]_{N_d \times N_d}$
  for $c = 1 : N_p$ **do**
    $e_{i,j} \leftarrow$ Choose the edge of E with the largest weight
    $k_{i,j} \leftarrow e_{i,j}$
    $G \leftarrow$ Eliminate vertices i and j from G
  end for
  Output: K

7.4.2 Proposed Device Pairing Algorithm

Here, we propose a heuristic approach to solve the device pairing sub-problem, which nominates maximum $N_p = T_{max}/T_{ts}$ edges of the graph iteratively for the NOMA transmissions. The algorithm works as follows. It iterates N_p times, where at each iteration, the edge with the largest weight is chosen. The devices corresponding to the vertices of that edge are added as a pair to the NOMA-pairing set, while they are eliminated from the graph. The algorithm terminates once either N_p pairs are nominated or no more edges with non-zero weights is remained. The result of the algorithm is presented by the matrix K with dimension of $N_d \times N_d$, where $k_{d,d'}$ is set to the weight of $e_{d,d'}$ if the pair is in the NOMA-pairing set, otherwise it is set to 0. Note that the reason to choose maximum N_p edges is the that length of the NOMA segment is restricted to N_p time-slots. It has been proved that the weight of the matching obtained by this algorithm is at least half of the weight of the maximum weight matching [8].

7.5 Scheduling Algorithm for NERA Scheme

In this section, given the device pairing graph obtained from Algorithm 7, we obtain the optimal selection of NOMA pairs along with p_d values for RA devices.

To solve the optimization problem, first we simplify the RA throughput by using approximations. Assuming that the network consists of a large number of devices, we have

$$\rho_d = \frac{y_d}{\prod_{d \in \mathcal{D}} (1 + y_d) - t'} \approx \frac{y_d}{\prod_{d \in \mathcal{D}} (1 + y_d)}$$

$$\approx y_d \left(1 - \sum_{d \in \mathcal{D}} y_d\right). \tag{7.3}$$

Substituting the above approximation into (7.2), we reach to the following optimization problem.

$$\max_{X^n,Y} \sum_{d\in\mathcal{D}} \left[\sum_{d'\in\mathcal{D}} \theta_d x_{d,d'}^n + y_d \left(1 - \sum_{d\in\mathcal{D}} y_d\right)\left(T_f - T_{ts} \sum_{d'\in\mathcal{D}} \sum_{d\in\mathcal{D}} x_{d,d'}^n\right)\right], \quad (7.4)$$

subject to:

C7.2.4–C7.2.7

The optimization problem in (7.4) is still non-convex. However, it can be solved by using an iterative algorithm, in which at each iteration the problem is decomposed into two sub-problems: (1) NDFA-ODFA scheduling, (2) p-probability derivation. More specifically, the whole algorithm consists of variable initialization followed by the iterative phase. In the iterative phase, first, for fixed values of Y, optimal X^n are derived and then in the next step, for the given X^n, optimal Y are obtained. The iterations continue until the results converge, which can be mathematically stated as

$$\|X^{n*}(t) - X^{n*}(t-1)\| \le \varepsilon_1, \text{ and}$$
$$\|Y^*(t) - Y^*(t-1)\| \le \varepsilon_2.$$

The procedure of finding optimal X^n and Y by using the proposed iterative algorithm can be expressed as

$$\underbrace{X^n(0) \to Y(0)}_{\text{Initialization}} \to \dots \underbrace{X^{n*}(t) \to Y^*(t)}_{\text{Iteration } t} \to \dots \to \underbrace{X^{n*} \to Y^*}_{\text{Optimal solution}}.$$

In the following, we discuss how to solve each of these sub-problems.

7.5.1 NDFA-ODFA Scheduling Sub-problem

This sub-problem takes the device pairing set obtained from Algorithm 7 and determines the selected pairs for the assignment-based NOMA or single devices chosen for the assignment-based OMA scheme. Here, we explain how the results of Algorithm 7 are represented in the NDFA-ODFA scheduling formulation.

Consider the matrix $F_{N_d \times N_d}$ with each element $f_{d,d'}$ defined as

$$f_{d,d'} = \begin{cases} 0.5 & \text{if } k_{d,d'} > 0 \ \& \ d \ne d', \\ 1 & \text{if } k_{d,d'} > 0 \ \& \ d = d', \\ 0 & \text{otherwise.} \end{cases} \quad (7.5)$$

The NDFA-ODFA scheduling sub-problem is expressed as

$$\max_{X^n} \sum_{d\in\mathcal{D}}\left[\sum_{d'\in\mathcal{D}}\theta_d f_{d,d'}x^n_{d,d'} + y_d(1-\sum_{d\in\mathcal{D}}y_d)(T_f-T_{ts}\sum_{d'\in\mathcal{D}}\sum_{d\in\mathcal{D}}f_{d,d'}x^n_{d,d'})\right], \quad (7.6)$$

subject to:

C7.6.1: $T_{ts}\sum_{d'\in\mathcal{D}}\sum_{d\in\mathcal{D}}f_{d,d'}x^n_{d,d'} \le T_{\max}$,

C7.6.2: $x^n_{d,d'} = x_{d',d}$, $\{\forall(d,d')|f_{d,d'}>0\}$,

C7.6.3: $\sum_{d'\in\mathcal{D}}x^n_{d,d'} \le 1, \forall d\in\mathcal{D}$,

C7.6.4: $x^n_{d,d'} \ge 0, \forall d\in\mathcal{D}, \forall d'\in\mathcal{D}$.

The above optimization problem is linear and can be solved by existing techniques.

7.5.2   p-Probability Derivation Sub-problem

For the fixed X^n, the probability derivation sub-problem is defined as

$$\max_{Y} \sum_{d\in\mathcal{D}}y_d(1-\sum_{d\in\mathcal{D}}y_d), \quad (7.7)$$

subject to:

C7.7.1: $(1+\theta_d)y_d \le \theta_d, \forall d\in\mathcal{D}$,

C7.7.2: $y_d\sum_{d'\in\mathcal{D}}x^n_{d,d'} = 0, \forall d\in\mathcal{D}$.

Notably, in the above optimization problem, the terms containing X^n are omitted in the objective function, while the resulting optimization is equivalent to the original problem.

The optimization problem in (7.7) is convex and hence the optimal solution can be derived.

7.5.3   Reconfigurable Access Scheme

In order to show how NERA can enhance the network performance leveraging NOMA, we compare its performance with the reconfigurable access (RCA) scheme, which is similar to the proposed NERA scheme, however it restricts the number of simultaneous transmissions in each slot to one. More specifically, in this scheme, at each frame, the following optimization problem should be solved.

Algorithm 8 NDFA-ODFA scheduling

Input: $F, \Theta, T_{\max}$
Initialization: Set initial value to (X, Y)
repeat
    $(X', Y') \leftarrow (X, Y)$
    Step 1: Find X

  • $X \leftarrow$ Solve the optimization problem (7.6) for fixed Y

    Step 2: Find Y

  • $Y \leftarrow$ Solve the optimization problem (7.7) for fixed X

until $|(X', Y') - (X, Y)| < \varsigma'$
$p_d \leftarrow \frac{y_d}{\theta d_s (1 + y_d)}$
Set $x_d = 1$ if it is in the sum(X) highest value of X, otherwise set $x_d = 0$
Output: X, P

$$\max_{X,Y} \sum_{d \in \mathcal{D}} \left[\theta_d x_d + \rho_d (T_f - T_{ts} \sum_{d \in \mathcal{D}} x_d) \right] \qquad (7.8)$$

subject to:

C7.8.1: $x_d y_d = 0, \qquad \forall d \in \mathcal{D},$

C7.8.2: $T_{ts} \sum_{d \in \mathcal{D}} x_d \leq T_{\max},$

C7.8.3: $\dfrac{y_d}{\theta_d (1 + y_d)} \leq 1, \qquad \forall d \in \mathcal{D},$

C7.8.4: $x_d \in \{0, 1\}, \qquad \forall d \in \mathcal{D}.$

In order to solve this problem, in a similar approach to the one for solving problem (7.2), we first apply the approximation (7.3) for ρ_d, and then we divide the problem into two sub-problems. In the first sub-problem, we obtain X for fixed Y, where the optimization problem is written as

$$\max_{X} \sum_{d \in \mathcal{D}} \theta_d x_d + y_d (1 - \sum_{d \in \mathcal{D}} y_d)(T_f - \sum_{d \in \mathcal{D}} x_d), \qquad (7.9)$$

subject to:

C7.9.1: $\log_2 \left(1 + \dfrac{p_t y_d}{n_d} \right) \geq R x_d, \ \forall d \in \mathcal{D},$

C7.9.2: $T_{ts} \sum_{d \in \mathcal{D}} x_d \leq T_{\max},$

C7.9.3: $x_d \leq 1, \forall d \in \mathcal{D}.$

In the second sub-problem, we obtain Y, for fixed X.

$$\max_{Y} \sum_{d \in \mathcal{D}} y_d \left(1 - \sum_{d \in \mathcal{D}} y_d\right), \tag{7.10}$$

subject to:

C7.10.1: $(1 + \theta_d) y_d \le \theta_d, \ \forall d \in \mathcal{D},$

C7.10.2: $x_d y_d = 0, \forall d \in \mathcal{D}.$

These sub-problems are solved sequentially over multiple rounds until the results of two last rounds converge.

7.5.4 Computational Complexity

In this sub-section, we analyze the computational complexity of the proposed algorithm. As the proposed algorithm is iterative, its computational complexity is the product of the number of iterations and complexity of each iteration.

Let first discuss the complexity of each iteration. In each iteration, two sub-problems should be solved. Thus, the computational complexity of each iteration that of these two sub-problems. The NDFA-ODFA scheduling sub-problem is a linear problem with $N_d^2 + N_d$ variables and at most $N_d^2 + N_d + N_p + 1$ constraints, which can be efficiently solved.

The second sub-problem for p-probability derivation is a convex optimization problem with N_d variables and $2N_d$ constraints, which can be solved in an efficient manner due to the low number of constraints.

Note that the number of iterations cannot be analytically studied. However, simulation results can be conducted to show its order of computational complexity.

7.6 Performance Evaluation

The results are produced in Matlab, and CVX is used for solving the optimization problems. We assume that the network consists of 300 devices, each with a packet arrival probability randomly chosen from the uniform distribution over $(0, 1)$. Devices are randomly located (with uniform distribution) in the coverage area of the AP. Furthermore, we set $N_{ts} = 100$, $T_{max} = 80$ time-slots, and $T_{ts} = 12$ time units. We also consider $\sigma^2 = 0$ and $Q_{max} = 4$. The results are obtained for 5 runs, each run consists of 100 frames.

To show the gain of the proposed scheme, we compare the results of the RCA scheme, with the same parameter values used in the NERA scheme. As a performance metric, we use normalized throughput. Here, the normalized throughput is

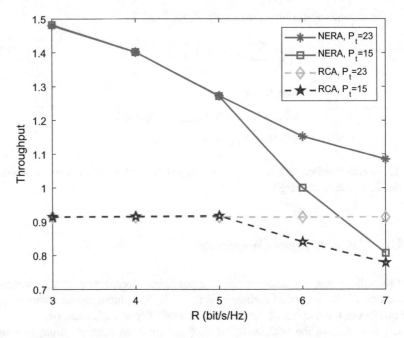

Fig. 7.2 Throughput versus R

defined as the average number of successfully transmitted packets per frame divided by the maximum throughput that can be achieved in an OMA scheme. Note that in OMA, the normalized throughput is not greater than one; however, as NOMA allows multiple devices to simultaneously transmit over the same time-slot, the normalized throughput can exceed one for NERA.

Figure 7.2 shows the normalized throughput versus different values of R for $p_t = 15$ and 23 dBm. As seen in this figure, by increasing R the performance of the NERA scheme decreases. The reason is that a smaller number of devices can be paired for NOMA transmissions when R is larger. Furthermore, as seen for the lower values of R, the same normalized throughput is achieved for both $p_t = 15$ and 23 dBm. However, after some point, the performance degradation for $p_t = 15$ dBm is higher. The reason is that, satisfying a larger rate requirement needs higher transmission powers. Therefore, with $p_t = 23$ dBm, higher throughput can be achieved. Moreover, the proposed NERA scheme outperforms the RCA scheme, and the performance gap is larger for lower rates.

Furthermore, in Figs. 7.3 and 7.4, we plot the average number of devices transmitting in the NDFA, ODFA and RA. By increasing R, the number of devices transmitting in NOMA decreases, while the number of devices transmitting in ODFA and RA increases. In fact, these plots explain why by increasing R, the performance gap between NERA and RCA scheme decreases.

Figure 7.5 shows the normalized throughput versus different values of p_t for two values of $R = 3$ and 5 bits/s/Hz. Evidently, NERA outperforms the RCA scheme

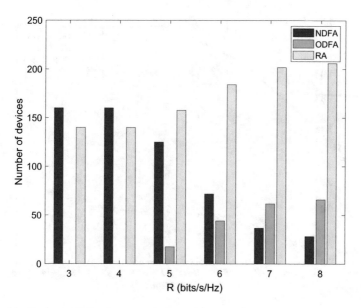

Fig. 7.3 Distribution in NERA for $p_t = 23$ dBm

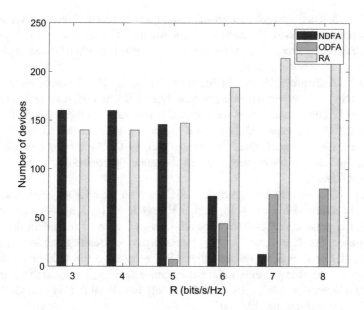

Fig. 7.4 Distribution in NERA for $p_t = 15$ dBm

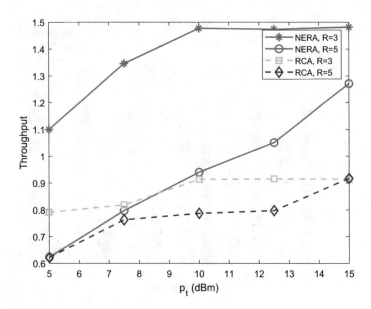

Fig. 7.5 Throughput versus p_t

when $R = 3$, since a larger number of devices can be paired. However, for $R = 5$, the performance of NERA is the same or close to the RCA scheme for lower powers, but it improves by increasing the power. The reason is that when the required rate is large, the pairing is not feasible.

Figure 7.6 illustrates the results for $p_t = 15$ dBm for $R = 3$ and 5. As observed, by increasing N_d the performance gap between NERA and RCA increases for both values of R. The reason is that more traffic is generated in the network, therefore devices with larger expected throughput are paired together. Moreover, the RCA throughput remains unchanged for $N_d \geq 200$, as ODFA throughput reaches to its capacity and RA throughput remains the same by controlling p values of p-persistent CSMA.

Figure 7.7 demonstrates the effect of SIC error on the NERA performance. The results are obtained for $\sigma^2 \in \{0, 0.05, 0.1\}$ and $R = \{1, 2, 3\}$. As observed, for $R = 1$, the presence of SIC error for all values of σ^2 does not affect the NERA performance. For $R = 2$, and low p_t, increasing σ^2 causes degradation on NERA throughput. For $R = 3$, in the presence of SIC error for $\sigma^2 = 0.05$ or 0.1, pairing is impossible, resulting in the similar performance as RCA. Thus, for this setting, in the case of imperfect SIC, using NOMA is only beneficial if R is low. Otherwise, no additional gain can be obtained.

Finally, Fig. 7.8 illustrates the average number of iterations that takes for NERA and RCA schemes to converge for the same setting of Fig. 7.6. As seen, both algorithms converge after a few iterations indicating a low computational complexity. Furthermore, for NERA, the number of iterations is larger when $R = 5$ compared to the case of $R = 3$.

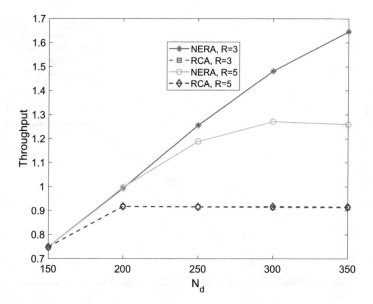

Fig. 7.6 Throughput versus N_d

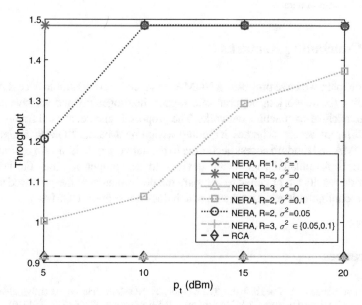

Fig. 7.7 Throughput versus p_t for different σ^2

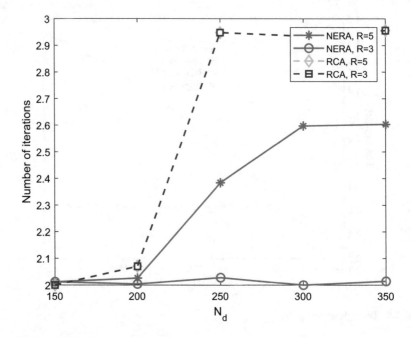

Fig. 7.8 Number of iterations

7.7   Concluding Remarks

In this chapter, we have proposed a NOMA-enhanced reconfigurable access scheme to increase the network throughput, and support heterogeneity and massive connectivity in machine-to-machine networks. The proposed scheme, at each frame, adopts three different access schemes including assignment-based NOMA, assignment-based OMA and random access according to the network condition. An optimization problem is formulated to allocate devices to the proper regime. Furthermore, simulation results are obtained to compare the performance of the proposed scheme with a reconfigurable access scheme which does not support NOMA.

References

1. A. Dalili Shoaei, M. Derakhshani, T. Le-Ngoc, A NOMA-enhanced reconfigurable access scheme with device pairing for M2M networks. IEEE Access **7**, 32266–32275 (2019)
2. A. Dalili Shoaei, M. Derakhshani, T. Le-Ngoc, A reconfigurable NOMA scheme for machine-to-machine networks, in *IEEE International Conference on Communication (ICC)*, Shanghai (2019)
3. Z. Wu, K. Lu, C. Jiang, X. Shao, Comprehensive study and comparison on 5G NOMA schemes. IEEE Access **6**, 18511–18519 (2018)

4. S. Lim, K. Ko, Non-orthogonal multiple access (NOMA) to enhance capacity in 5G. Int. J. Contents **11**(4), 38–43 (2015)
5. J. Edmonds, Paths, trees, and flowers. Can. J. Math. **17**(3), 449–467 (1965)
6. H.N. Gabow, Data structures for weighted matching and nearest common ancestors with linking, in *ACM-SIAM Symposium on Discrete Algorithms*, San Francisco (1990)
7. M. Wattenhofer, R. Wattenhofer, Fast and simple algorithms for weighted perfect matching. Electron. Notes Discrete Math. **17**, 285–291 (2004)
8. D. Avis, A survey of heuristics for the weighted matching problem. Networks **13**(4), 475–493 (1983)

Chapter 8
A Distributed Contention-Resolution Self-Organizing TDMA Scheme for MTC

8.1 Introduction

M2M networks are comprised of a variety of applications, which are heterogeneous in terms of QoS requirements. Furthermore, the size of data packets is small in MTC. These two features of M2M communications lead to the demand for a robust and resilient access scheme which can support diverse QoS requirements at low signaling overhead. In the previous chapters, the idea of using a reconfigurable access scheme in which the channel access time is divided into two segments is proposed. However, in the DFA segment of this scheme, devices are scheduled by the AP which causes signaling overhead. In order to minimize this signaling overhead, specifically for scenarios that data packets are too small, distributed access schemes can be deployed.

One potential distributed access scheme that has low signaling overhead and can converge to collision-free data transmission is pseudo-TDMA (PTDMA) [1–5]. In PTDMA, each active device starts its transmission by choosing random back-off times as in CSMA. But, after a device successfully transmits once and receives the relevant acknowledgement, it picks a fixed deterministic back-off value and switches to periodic transmission as in TDMA. After any collision, the device switches back to random back-off. Although periodic transmission can provide reservation guarantees by having fixed slots and a frame-length by which the slot allocation is repeated, it has a drawback of being inflexible and cannot adapt to heterogeneous traffic. Thus, a MAC protocol that employs PTDMA but circumvents its inflexibility problem is necessary and will be the focus of this chapter.

In order to identify and also quantify both potentials and shortcomings of PTDMA in QoS support, in this chapter [6, 7], we first investigate the *effective capacity* (EC) of PTDMA in comparison with CSMA. This study evaluates the statistical QoS performance of PTDMA under a variety of traffic conditions. EC has been proposed in [8] as a QoS-aware metric that determines the maximum constant

© Springer Nature Switzerland AG 2020

T. Le-Ngoc, A. Dalili Shoaei, *Learning-Based Reconfigurable Multiple Access Schemes for Virtualized MTC Networks*, Wireless Networks, https://doi.org/10.1007/978-3-030-60382-3_8

arrival rate that can be supported by a network, while satisfying a target statistical delay requirement. Our studies show that PTDMA can improve EC in *saturated* traffic scenarios compared with CSMA by reducing the collision probability.

However, when traffic is *unsaturated* in the network, our studies reveal that CSMA could provide better EC performance. Such unsaturated traffic condition results in short-term random fluctuations in the number of active devices (denoted by N_a), that have packets in their queues and are contending for the channel. This variation in N_a is not catered for in the PTDMA frame structure. In other words, PTDMA parameters such as the time-slot size and the frame-length are not adjusted based on N_a, while CSMA is flexible to the traffic demand and the number of active devices owing to its opportunistic behavior. Consequently, in an unsaturated traffic scenario, PTDMA suffers from underutilization that may lead to high delays. The reason is that a portion of the channel might be left unoccupied even though there are devices waiting to send a packet.

In order to overcome this problem, we propose a self-organizing TDMA (SO-TDMA) protocol. In this protocol, devices can operate in a distributed and asynchronous manner by carrier sensing, but eventually in an efficient and opportunistic TDMA manner. More specifically, each device initiates its transmission through CSMA and then switches to periodic transmission as in PTDMA. But, different from PTDMA, the wireless channel frame structure in SO-TDMA is adaptable to the changing traffic and channel conditions.

In particular, a distributed learning-based MAC algorithm is developed, in which each device independently adapts its transmission length to the optimal values over time by learning the number of active devices based on locally available information. This fully distributed SO-TDMA protocol eliminates the need for any central coordination that would suffer from scalability issues or any information exchange that would degrade throughput due to additional overhead. The process of transmission length adaptation of SO-TDMA will be analytically derived from the network congestion control problem, where each device independently adapts its own transmission rate to avoid any congestion in the bottleneck link due to the limited capacity.

The convergence behavior of proposed SO-TDMA is analytically studied for two phases of the proposed algorithm. First, the required time to pass the initial CSMA phase (modeled as a Markov chain with one absorbing state) is studied. It is proved that the expected time for transition to a periodic transmission phase is only linearly increasing with the network size. Second, for the periodic transmission phase, it is proved that the proposed additive-increase-multiplicative-decrease (AIMD) time-slot adaptation algorithm for SO-TDMA converges to fairness.

Simulation results are obtained to evaluate the performance of the proposed SO-TDMA algorithm in terms of system throughput, fairness, collision probability, and effective capacity for QoS. They reveal that SO-TDMA can be a good MAC candidate for QoS provisioning in MTC for better EC as compared with CSMA and PTDMA in both saturated and unsaturated traffic scenarios.

The rest of this chapter is organized as follows. Section 8.2 describes the system model for all the MAC protocols considered as well as a brief review of PTDMA and other related works. After details of the proposed SO-TDMA protocol given in Sect. 8.3, Sect. 8.4 studies the convergence properties of SO-TDMA. This is followed by the performance evaluation and benchmarking results in Sect. 8.5. Finally Sect. 8.6 presents concluding remarks.

8.2 System Model

We consider a network comprised of a single AP and a set of devices $\mathcal{D} = \{1, \cdots, N_d\}$ sharing a communication channel, where the average arrival bit rate of each device d is μ_d as shown in Fig. 8.1. We study two different types of arrival traffic in this work, including constant bit rate (CBR) and Poisson arrival. Suppose Q_d denotes the queue length of device d, i.e., the number of backlogged packets in its queue. Accordingly, the number of active devices (i.e., N_a) is defined as the number of devices with $Q_d > 0$ whose instantaneous SNR is also above the minimum required threshold for transmission.

We assume a slotted transmission, where T denotes the length of the smallest unit of time which is a back-off time-slot. Subsequently, the length of a single transmission opportunity of device d is represented by $T_{s,d}$, which is defined in terms of the number of back-off time-slots. Furthermore, for device d, the required number of back-off time-slots to transmit one packet can be calculated as $S_{req,d} = P_s/(R_d T)$, where P_s is the fixed packet size and R_d is the transmission data rate of device d.

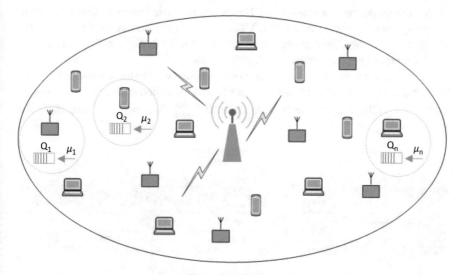

Fig. 8.1 Network model

A block fading channel model is assumed in which each device's channel gain remains constant for a block of time equal to the channel coherence time T_c, while it independently and asynchronously varies from one block to another. The transmission data rate of device d, R_d, is determined based on the instantaneous channel power gain, which is reported back to each device by the AP.

Based on this information, a limited number of transmission modes, K, corresponding to certain transmission rates, is selected in order to guarantee a minimum packet error rate (PER) for each device. Let g_d be the channel power gain between the device d and the AP, the instantaneous SNR is given as

$$SNR_d = Pg_d/\sigma^2, \tag{8.1}$$

where P is the transmit power of the device and σ^2 represents the noise power. A transmission mode $k \in \{1, \ldots, K\}$ corresponding to the channel rate R_k requires a minimum SNR threshold η_k where $\eta_1 < \eta_2 < \cdots < \eta_K$. Therefore, the selected channel rate R_k for transmission of packets corresponds to the η_k such that $SNR_d \geq \eta_k$. If $SNR_d < \eta_1$ no transmission takes place. The basic concept of PTDMA is illustrated in Fig. 8.2. PTDMA starts with CSMA and then switches to a *periodic transmission phase* with a frame-length equal to T_f. There are two important parameters related to PTDMA: the frame-length, T_f, and the transmission time-slot size, $T_{s,d}$. Both are vital to its performance as they directly impact the channel utilization, throughput, packet delay and fairness.

Since PTDMA has fixed parameters that define the frame structure, due to the dynamic nature of the channel and traffic arrival, it is often the case that some portion of the channel is left unused, adversely affecting channel utilization. This is due to two possible scenarios: (1) The dynamic queue length of devices leads to empty queues in several of them, (2) The channel gain between a device and the AP falls below the minimum threshold, prohibiting the device from transmitting or rendering its signal too weak to be decoded at the AP.

A device that is not in the two scenarios mentioned above (has a non-empty queue and channel gain equal to or over the thresholds) are called *active devices*. In order to achieve 100% channel utilization, we need $\sum_{d=1}^{N_a} T_{s,d} = T_f$, where N_a is the

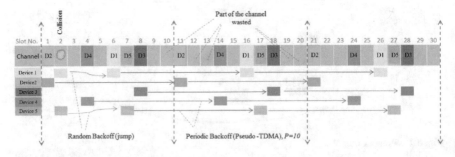

Fig. 8.2 Illustration of PTDMA operation

number of active devices. Ideally, $T_{s,d}$ should instantaneously be updated to $T_{s,d} = T_f/N_a$. In the rest of this chapter, we refer to PTDMA with $T_{s,d} = T_f/N_a$ as "Ideal-PTDMA" and implement it to represent an upper bound for PTDMA performance. However, in reality, since the information of N_a is hard to be available to every device in the network, in PTDMA, $T_{s,d}$ is generally set to $T_{s,d} = T_f/N$, where N_d is the total number of devices and its value is assumed to be known by every device.

An enhanced version of PTDMA, called period-controlled MAC (PCMAC), is proposed in [9], in which the frame-length is not fixed and adapts to the number of idle slots, while maintaining the relative position of each device in each frame-length. Since the convergence of the PTDMA is very sensitive to the frame-length, changing it can be problematic considering fluctuating heterogeneous traffic demand and random channel gains. Moreover, since T_f determines the delay between successive transmissions, its value can significantly impact QoS for delay sensitive applications and changing it could not be the best option.

Thus, in the following, we propose SO-TDMA, an alternative of PTDMA with adjustable time-slot sizes and fixed frame-length. With this approach, we can still have the adaptability with the network load and channel gains as in PCMAC, while securing reliable convergence and better EC performance. In the proposed SO-TDMA, the transmission length of each device is distributively optimized during the *periodic transmission phase* in order to maximize channel utilization and fairness. Through AIMD control, the value of transmission time-slot size $T_{s,d}$ converges from an initial value of T_0 to $\frac{T_f}{N_a}$ and aims to closely follow this value. Therefore, this protocol allows devices to optimize the pseudo-frame structure according to the changing channel and traffic conditions, repeating the cycle of *observe*, *decide*, *act* and *learn* every pseudo-frame while approaching the performance of the Ideal-PTDMA. Since for the Ideal-PTDMA algorithm the instantaneous value of N_a is assumed to be available, its performance is optimal. In order to verify the benefits of the proposed SO-TDMA protocol, through numerical results, its performance is compared with CSMA, PTDMA, Ideal-PTDMA, and PCMAC as benchmarks.

8.3   Proposed Self-Organizing TDMA (SO-TDMA)

Here, we present in detail the SO-TDMA protocol that aims to improve the spectrum sharing efficiency in terms of channel utilization, fairness among devices, and QoS in terms of packet delay. The key advantage of the proposed SO-TDMA is the ability to dynamically adapt the transmission time-slot of each device, $T_{s,d}(f_d)$, to changing traffic and channel conditions, where f_d represents the pseudo-frame index for device d. Thus, this protocol aims to cater as many devices on the common channel as possible in their respective pseudo-frames, while minimizing the number of collisions as well as the wastage of resources. The access mechanism in SO-TDMA consists of two phases for each device: (1) *initial access phase* where each device tries to access the channel via the random back-off procedure of CSMA; (2)

periodic transmission phase where each device independently attempts to shape the network pseudo-frame structure, while maximizing the channel utilization and the fair slot allocation among devices.

8.3.1   Initial Access Phase

In SO-TDMA, the *initial access phase* for a device is the same as in CSMA, where the exponential random back-off procedure is executed. More specifically, a device with a packet to transmit should monitor the channel before transmission for a period of time called distributed inter frame space (DIFS). If the channel is sensed idle, the device will transmit. Otherwise, it continues monitoring until the channel is measured idle for a duration of DIFS and backs off for a random period of time, uniformly chosen from a range of $[0, w - 1]$, where w is known as the contention window given in terms of back-off time-slots. Initially, w is set to the minimum contention window size $CW_{\min}$. In case of any failure attempt, the value of w is doubled but not exceeding the maximum contention window size $CW_{\max}$ [10].

The back-off counter is decremented and a device transmits when it reaches zero. While counting down, if the channel is sensed busy, the device will freeze its back-off counter and continue decrementing once channel becomes idle again. For each transmission opportunity, the device can transmit multiple back-to-back packets for a fixed period of time equal to $T_{s,d}$. Once the packet is received successfully, the AP waits for a period of time called short inter-frame space (SIFS) and then sends an acknowledgment (ACK) for the successful reception.

In the *initial access phase*, the transmission time-slot duration is fixed to $T_{s,d}(f_d) = T_0$ for all devices. After a device grabs the channel for the first time using CSMA, a countdown timer of length T_f is initiated at the start of its transmission. In this phase, if the device could empty its queue during any transmission (i.e., $Q_d = 0$), the timer is reset to T_f again. In case the timer expires and the device still has some packets in its queue (i.e., $Q_d > 0$), the device would enter the *periodic transmission phase* (which is presented in Algorithm 9) until Q_d becomes 0 again.

8.3.2   Periodic Transmission Phase

In the *periodic transmission phase*, all devices periodically transmit with the same back-off value equal to the *pseudo-frame length* T_f. At the end of each device's respective periodic back-off, it senses the channel and starts transmission if the channel is sensed idle with the transmission time-slot size $T_{s,d}(f_d + 1)$. The value of $T_{s,d}(f_d + 1)$ is updated according to Algorithm 9, which will fully be described later. In case the channel is not idle, the device *does not revert* to the initial access phase but tries to find another back-off value to resume the *periodic transmission phase* by the following manner. First, the device continues sensing the channel until it becomes idle again. Then, the device carries out an additional random back-off procedure before returning to the periodic transmission phase. This additional random back-off is necessary to avoid collisions between two devices with periodic back-off already expired and waiting to transmit as the channel became free. In the following section, the transmission time-slot adaptation algorithm is explained.

8.3.2.1 AIMD Time-Slot Adaptation Algorithm

Suppose that all devices picked their locations in the frame through CSMA in the initial phase and started to transmit in the *periodic transmission phase*. The process of adjusting the transmission time-slot size can be modeled as a network congestion control problem, in which different devices independently adapt their transmission rates to avoid any congestion in the bottleneck link due to the limited capacity [11–13].

Let define the *utilization rate* of device d as $r_d = \frac{T_{s,d}}{T_f}$ where $0 \leq r_d \leq 1$. The rate allocation can be modeled as an optimization problem to maximize the channel utilization and to achieve fairness as

$$\max_{r \geq 0} \sum_{d=1}^{N_d} U_d(r_d), \quad \text{subject to,}$$

$$\sum_{d=1}^{N_d} r_d \leq 1 \tag{8.2}$$

where $r = [r_1, \ldots, r_{N_d}]$. $U_d(r_d)$ represents the utility earned by device d with allocated r_d. The linear flow constraint also states that the sum of time-slot sizes should be kept smaller that the frame size.

Assuming $U_d(r_d) = \log(r_d)$, in [12], it has been proven that a utilization rate control, based on a system of differential equations is as

$$\frac{\partial}{\partial t} r_d(t) = \kappa \left(1 - r_d(t)\beta(t)\right) \tag{8.3}$$

where

$$\beta(t) = h\left(\sum_{d'=1}^{N_d} r_{d'}(t)\right), \tag{8.4}$$

will converge to a unique stable point, which satisfies *proportional fairness*. In the system (8.3)–(8.4), $\kappa > 0$ is a constant and $h(x)$ is a non-negative, continuous, and increasing function of x. A feasible vector $r^* = [r_1^*, \ldots, r_{N_d}^*]$ is *proportionally fair* if

$$\sum_{d=1}^{N_d} \frac{r_d - r_d^*}{r_d^*} \leq 0 \quad \Longleftrightarrow \quad \frac{1}{N_d} \sum_{d=1}^{N_d} \frac{r_d}{r_d^*} \leq 1 \tag{8.5}$$

for any other feasible vector r, i.e., the aggregate of proportional changes is zero or negative.

Assuming $h(x) = [x - 1 + \epsilon]^+ / \epsilon^2$ where $[x]^+ = \max(x, 0)$, it has been shown that (8.3) converges to the optimal solution of (8.2) when $\epsilon \to 0$.

$\beta(t) = \left[\sum_{d'=1}^{N} r_{d'}(t) - 1 + \epsilon \right]^+ / \epsilon^2$ is a pricing function to control the rate of change of $r_d(t)$. When the system is congested, $\beta(t)$ is large (for small ϵ) and the rate of device d will be decreased proportional to its rate as $1 - r_d(t)\beta(t) < 0$. Accordingly, supposing $\Delta t = T_f$, the utilization rate control can be written as

$$r_d(f_d + 1) \approx \tag{8.6}$$

$$r_d(f_d) + \kappa T_f \left(1 - r_d(f_d) \left[\sum_{d'=1}^{N_d} r_{d'}(f_d) - 1 + \epsilon \right]^+ / \epsilon^2 \right).$$

Consequently, (8.6) becomes

$$r_d(f_d + 1) \approx r_d(f_d) + \tag{8.7}$$

$$\begin{cases} \kappa T_f & \text{if } \sum_{d'=1}^{N_d} r_{d'}(f_d) \leq 1 - \epsilon \\ \kappa T_f \left(1 - r_d(f_d) \left[\sum_{d'=1}^{N_d} r_{d'}(f_d) - 1 + \epsilon \right] / \epsilon^2 \right) & \text{otherwise} \end{cases}.$$

Let define $R_I = \kappa T_f$ and $R_D = \kappa T_f / \epsilon$. Moreover, we define the *expected* idle fraction of the frame f_d as

$$y(f_d) = 1 - \sum_{d'=1}^{N_d} r_{d'}(f_d). \tag{8.8}$$

According to these definitions, (8.7) becomes

$$r_d(f_d + 1) \approx \tag{8.9}$$

$$\begin{cases} r_d(f_d) + R_I & \text{if } y(f_d) \geq \epsilon \\ r_d(f_d) + R_I - R_D r_d(f_d)[1 - y(f_d)/\epsilon] & \text{if } y(f_d) \leq \epsilon \end{cases}.$$

This utilization rate control is kind of an *additive-increase-multiplicative-decrease (AIMD)* algorithm, in which if the expected idle fraction of a current frame is larger than a threshold, ϵ, then the utilization rate of each device will be increased by adding a constant to its previous demand. However, if the current frame is mostly busy and the expected idle fraction is smaller than ϵ, then the utilization rate of each device will be reduced and the decrease is multiplicative proportional to the previous utilization rate. Here, ϵ can be interpreted as

$$\epsilon = I_{\text{th}}/T_f, \tag{8.10}$$

where I_{th} is defined as the target number of idle back-off units in each frame, in order to provide the non-active devices a non-zero access chance to the channel.

To transform the utilization rate control in (8.9) into a time-slot adaptation, setting $W_I = R_I T_f$ and $W_D = R_D$, we have

$$T_{s,d}(f_d + 1) \approx \tag{8.11}$$

$$\begin{cases} T_{s,d}(f_d) + W_I & \text{if} \quad J(f_d) \geq I_{th} \\ T_{s,d}(f_d) + W_I - W_D T_{s,d}(f_d)(1 - \frac{J(f_d)}{I_{th}}) & \text{if } J(f_d) \leq I_{th} \end{cases}$$

where

$$J(f_d) = T_f - \sum_{d'=1}^{N_d} T_{s,d'}(f_d). \tag{8.12}$$

$J(f_d)$ gives an estimation of the network load, i.e., the lower J indicates the busier channel we would expect. However, in reality, it is hard to inform all devices about the time slot size of all other devices, so each device tries to measure the number of idle back-off units in its pseudo-frame as an estimation of $J(f_d)$. Nonetheless, this measurement is noisy due to collision, in other words, if we define $I_d(f_d)$ as the *measured* number of idle back-off units sensed by device d during the pseudo-frame f_d, we have

$$I_d(f_d) = J(f_d) + z_d(f_d), \tag{8.13}$$

where $z_d(f_d)$ represents the noise during the measurement by device d in pseudo-frame f_d. In addition to collisions that may cause an error in measuring the real number of idle time-slots, asynchronicity in measurements can cause a problem. In other words, since the start and end of the pseudo-frame for each device might be different, $I_d(f_d)$ measurements during the time period T_f after each transmission are asynchronous. However, assuming that all the devices can perfectly sense the channel status, i.e., busy or idle, a moving average $\bar{I}_d(f_d)$ on the number of idle time-slots enables all active devices to ultimately converge to a same value.

$$\bar{I}_d(f_d) = \alpha I_d(f_d) + (1 - \alpha)\bar{I}_d(f_d - 1), \ 0 < \alpha < 1 \tag{8.14}$$

Thus, we can update the control algorithm in (8.11) using a moving average of I_d as a measure for J, which can partly phase out the noise.

$$T_{s,d}(f_d + 1) = \tag{8.15}$$

$$\begin{cases} T_{s,d}(f_d) + W_I & \text{if} \quad \bar{I}_d(f_d) \geq I_{th} \\ T_{s,d}(f_d) + W_I - W_D T_{s,d}(f_d)(1 - \frac{\bar{I}_d(t)}{I_{th}}) & \text{if } \bar{I}_d(f_d) \leq I_{th} \end{cases}$$

Thus, using the time-slot adaption in (8.15), each device aims to ensure that there are at least I_{th} idle time-slots in all pseudo-frames. To reach this goal, each device adaptively adjusts its transmission time-slot size in the next frame, i.e., $T_{s,d}(f_d + 1)$, based on the information from the last frame $T_{s,d}(f_d)$ and $\bar{I}_d(f_d)$. When the frame is underutilized, i.e., $\bar{I}_d(f_d) > I_{th}$, $T_{s,d}$ will be increased by adding a constant (W_I) to the previous demand. However, when the frame is over-utilized $\bar{I}_d(f_d) < I_{th}$, the time-slot size will be decreased proportional to the previous size. The details of the proposed time-slot adaptation algorithm are presented in Algorithm 9 (Lines 8–12).

In Algorithm 9, it should be noted that T_{min} and T_{max} are considered as the lower and upper bounds on the value of $T_{s,d}$, respectively. In other words, at any frame, $T_{s,d}$ needs to be kept in a range as $T_{min} \leq T_{s,d} \leq T_{max}$. Thus, after adjusting $T_{s,d}$ based on AIMD technique in Lines 8–12, such enforcement is implemented in Line 13 as

$$T_{s,d}(f_d + 1) = \begin{cases} T_{max} & \text{if} \quad T_{s,d}(f_d + 1) > T_{max} \\ T_{min} & \text{if} \quad T_{s,d}(f_d + 1) < T_{min} \\ T_{s,d}(f_d + 1) & \text{Otherwise} \end{cases} . \tag{8.16}$$

T_{min} can be set to an arbitrary value so that the payload is still a significant portion of a transmission as compared to the overheads. However, T_{min} directly limits the maximum number of active devices in the fixed frame-length T_f to T_f/T_{min}. Moreover, we need to have $T_{max} \leq T_f - I_{th}$ in order to ensure at least I_{th} idle time-slots in any frame, even if there is a single active device.

Algorithm 9 SO-TDMA: AIMD time-slot adaptation at active device d

---
1: **Initialization:**
    Set T_{min}, T_{max}, and T_0
    Set $f_d = 0$, $T_{s,d}(0) = T_0$ and T_f
    Set I_{th}, $\bar{I}_d(0) = I_{th}$
2: **Periodic Transmission Phase**
3: **while** $Q_d(f_d) > 0$ **do**
4:     Measure $I_d(f_d)$
5:     Set $\bar{I}_d(f_d) = \alpha I_d(f_d) + (1 - \alpha)\bar{I}_d(f_d - 1)$
6:     **if** $\bar{I}_d(f_d) > I_{th}$ **then**
7:         $T_{s,d}(f_d + 1) = T_{s,d}(f_d) + W_I$
8:     **else if** $\bar{I}_d(f_d) < I_{th}$ **then**
9:         $T_{s,d}(f_d + 1) = T_{s,d}(f_d)(1 - W_D(1 - \frac{\bar{I}_d(t)}{I_{th}})) + W_I$
10:     **end if**
11:     $T_{s,d}(f_d + 1) = \max(\min(T_{s,d}(f_d + 1), T_{max}), T_{min})$
12:     $f_d = f_d + 1$
13:     **if** $Q_d(f_d) = 0$ **then**
14:         Return to Initial access phase
15:         $f_d = 0$, $T_{s,d}(0) = T_{start}$
16:     **end if**
17: **end while**

In this algorithm, the moving average parameter α determines the amount of stability in the algorithm (a higher value gives more weightage to the latest measure of idle-slots as opposed to previous values). Decreasing α increases stability but also increases the convergence time, while increasing α can reduce the convergence time up to a certain point beyond which it becomes too unstable to converge. Furthermore, I_{th} (the target threshold for the number of idle time-slots in one frame), defined to give access probability to new incoming devices, also faces a compromise. Increasing its value will make channel utilization poor as more back-off slots per frame are always kept idle but will reduce the initial channel access delay as the probability of finding idle slots for new devices becomes higher. Similarly, decreasing it will have the opposite impact.

This iterative process in Algorithm 9 runs independently and in a distributed manner on device d until $Q_d(f_d) = 0$. In a nutshell, Algorithm 9 ensures how devices converge to a common pseudo-frame structure without any information sharing, while achieving fairness at the pseudo-frame level (avoiding starvation of devices) and high channel utilization.

8.4 Convergence Analysis

In this section, we study convergence properties of the proposed SO-TDMA, including the required time to pass the initial access phase and convergence to fairness in the periodic transmission phase using AIMD time-slot adaptation.

8.4.1 Convergence Time: Initial Access Phase

In this subsection, we aim to study the required time for passing the initial phase and moving on to the periodic transmission phase. To this end, we first try to mathematically model the initial access phase of SO-TDMA. Then, using the developed model, we investigate the asymptotic behavior of convergence time with respect to the number of devices in the network.

To present a tractable model, we make a few assumptions. First, we assume all devices have always packets to transmit, which can represent the worst case scenario. Since all the devices start with a time-slot equal to T_{start}, we assume there are $M = \frac{T_f}{T_{\text{start}}}$ different positions available in a frame to be picked by each device. We also assume devices will choose their locations in a round-robin manner. More specifically, in each round, one device randomly picks (via CSMA) a location out of the M available ones. If the device chooses a location which already picked by others, a collision occurs. Thus, the two devices who have been involved in this collision need to keep trying to find appropriate locations.

With these assumptions, the initial access phase of SO-TDMA can be modeled as a discrete-time Markov chain with one absorbing state. Let assume that the state of the network (denoted by S) is represented by the number of devices, who have already accessed the channel successfully through CSMA and grabbed their locations in the frame. Thus, S belongs to a set $\mathcal{S} = \{0, \ldots, N_d\}$.

This procedure can be modeled as a discrete-time Markov chain since the probability of the network having N_d settled devices in the future only depends on the current number of devices in the network not the past. In other words, the initial access phase of SO-TDMA has the Markov property as

$$\Pr(S_{i+1} = N_d | S_0 = n_0, S_1 = n_1, \ldots, S_i = n_i)$$

$$= \Pr(S_{i+1} = N_d | S_i = n_i). \tag{8.17}$$

This Markov chain is a birth and death chain, except for the last state that represents the case when all devices settled down in the frame through CSMA. Thus, the state N_d is the absorbing state with $\Pr(S_{i+1} = N_d | S_i = N_d) = 1$.

Otherwise, if the network is at the state $S_i = s \in \{0, \ldots, N_d - 1\}$, the chance that a device, whose turn is to pick, chooses an idle location is $1 - \frac{s}{M}$. Thus, we have

$$\Pr(S_{i+1} = s + 1 | S_i = s) = 1 - \frac{s}{M}. \tag{8.18}$$

Otherwise, there is a chance of $\frac{s}{M}$ that the device chooses a location, which has been already picked by others. In this case, a collision occurs and the system state will go back to $s - 1$. Thus, we have

$$\Pr(S_{i+1} = s - 1 | S_i = s) = \frac{s}{M}. \tag{8.19}$$

In order to calculate the convergence time in the initial phase, we study the mean time to absorption in the underlying Markov chain. In other words, our objective is to find the expected number of visits in the transient states before being absorbed at $s = N_d$, when starting at $s = 0$.

Proposition 8.1 *The expected time to converge to a periodic transmission phase is linearly increasing with N_d when $N_d < \sqrt{M}$.*

Proof The transition matrix of the proposed Markov chain can be represented as

$$P = \begin{bmatrix} A & B \\ 0 & I_1 \end{bmatrix}, \tag{8.20}$$

where the $N_d \times N_d$ *absorbing matrix* A represents the probabilities of transitions among the transient states, the $N_d \times 1$ matrix B represents the probabilities to reach the absorbing state from other transient states in one step, 0 is an $1 \times N_d$ zero matrix, and I_1 is the 1×1 identity matrix. In this problem, the entries of B are 0, except for the last one $b_{N_d} = 1 - \frac{N_d - 1}{M}$, which represents the transition probability from $s = N_d - 1$ to the absorbing state $s = N_d$. In an absorbing Markov chain, it is well-known that the matrix of the expected number of visits to each state before absorption given the initial state is

$$L = [l_{ij}]_{N_d \times N_d} = (I_{N_d} - A)^{-1}, \tag{8.21}$$

where l_{ij} represents the expected number of times the chain is in state j, given that the chain started in state i [14]. Here we are looking for

$$\lambda_0 = \sum_{j=0}^{N_d-1} l_{1j}, \tag{8.22}$$

which is the expected number of visits in the transient states before reaching $s = N_d$ when starting at $s = 0$. In the proposed Markov chain, $I_{N_d} - A$ is a triangular matrix as

$$I_{dN} - A = \begin{bmatrix} 1 & b_1 & 0 & 0 & \ldots & 0 \\ c_1 & 1 & b_2 & 0 & \ldots & 0 \\ 0 & c_2 & 1 & b_3 & \ldots & 0 \\ & \ddots & \ddots & \ddots & & \\ 0 & 0 & \ldots & c_{N_d-2} & 1 & b_{N_d-1} \\ 0 & 0 & \ldots & 0 & c_{N_d-1} & 1 \end{bmatrix},$$

where $b_j = -1 + \frac{j-1}{M}$ and $c_j = -\frac{j}{M}$ for $j \in \{1, \ldots, N_d - 1\}$. The inversion of this tridiagonal matrix can be calculated numerically by

$$l_{1j} = (-1)^{1+j} b_1 b_2 \ldots b_{j-1} \frac{\Phi_j + 1}{\Theta_{N_d}}, \tag{8.23}$$

where Θ_i and Φ_i satisfy the recurrence relation as

$$\Theta_j = \Theta_{j-1} - b_{j-1} c_{j-1} \Theta_{j-2}, \tag{8.24}$$

$$\Phi_j = \Phi_{j+1} - b_j c_j \Phi_{j+2}, \tag{8.25}$$

with initial conditions $\Theta_0 = \Theta_1 = \Phi_{N_d} = \Phi_{N_d+1} = 1$ [15].

To find a closed-form approximation for λ_0, we compute an upper bound. Considering that $b_j c_j > 0$, from (8.25), it is clear that $\Phi_j \leq 1$. Thus, (8.23) can be bounded as

$$l_{1j} \leq \frac{(-1)^{1+j}}{\Theta_{N_d}} b_1 b_2 \ldots b_{j-1}. \tag{8.26}$$

Consequently,

$$\lambda_0 = \sum_{j=0}^{N_d-1} l_{ij} \leq \frac{1}{\Theta_{N_d}} \left[2 + \sum_{j=1}^{N_d-2} \left(1 - \frac{j}{M}\right) \right] \leq \frac{N_d}{\Theta_{N_d}}. \tag{8.27}$$

Since $0 < b_j c_j \leq -c_j$, (8.24) can be lower bounded by

$$\Theta_j \geq \Theta_{j-1} + c_{j-1}\Theta_{j-2} = \Theta_{j-1} - \frac{j-1}{M}\Theta_{j-2}. \tag{8.28}$$

Lemma 8.1 *The lower bound on the Θ_{N_d} can be derived as*

$$\Theta_{N_d} \geq 1 - \sum_{j=1}^{N_d-1} \frac{j}{M} = 1 - \frac{N_d(N_d-1)}{2M}. \tag{8.29}$$

Proof See Appendix.

Thus, considering (8.27) and *Lemma 8.1*, an upper bound can be obtained for λ_0 as

$$\lambda_0 \leq \frac{N_d}{1 - \frac{N_d(N_d-1)}{2M}}. \tag{8.30}$$

From (8.30), the mean time to absorption in the presented Markov chain starting at $s = 0$ is linearly increasing with N_d when $N_d < \sqrt{M}$.

Proposition 8.1 confirms that SO-TDMA passes the initial access phase in a time window that is linearly proportional to the network size and then settles in the periodic transmission phase.

8.4.2 Convergence to Fairness: Periodic Transmission Phase

In this subsection, we prove that the proposed AIMD time-slot adaptation algorithm in the periodic transmission phase of SO-TDMA converges to fairness. In other words, following Algorithm 9, devices can reach to a state, where each has an equal share of channel-access. The Jain's index [16] to evaluate fairness at pseudo-frame f_n can be calculated as

$$F(f_d) = \frac{1}{N_d} \frac{(\sum_{d=1}^{N_d} T_{s,d}(f_d))^2}{\sum_{d=1}^{N_d} T_{s,d}^2(f_d)}. \tag{8.31}$$

Proposition 8.2 *The fairness index monotonically converges to one, i.e.,*

$$\forall d \ F(f_d) < F(f_d + 1) \ and \ \lim_{d \to +\infty} F(f_d) = 1. \tag{8.32}$$

Proof In [11], it has been shown that

$$F(f_d + 1) = \tag{8.33}$$

$$F(f_d) + (1 - F(f_d))\left(1 - \frac{\sum_{d=1}^{N_d} T_{s,d}^2(f_d)}{\sum_{d=1}^{N_d} (C_d + T_{s,d}(f_d))^2}\right),$$

where, according to (8.11),

$$C_d = \begin{cases} W_I, & \text{if } J(f_d) \geq I_{th} \\ W_I \times \frac{1}{1-W_D(1-J(f_d)/I_{th})}, & \text{if } J(f_d) \leq I_{th} \end{cases}.$$

Since for $W_D < 1$, C_d is always positive, the second term in (8.33) will be positive and $F(f_d) < F(f_d + 1)$. As a result of the strict increase of $F(f_d)$ over f_d and $0 \leq F \leq 1$, it can be concluded that $\lim_{d \to +\infty} F(f_d) = 1$.

8.5   Illustrative Results

In this section, first, the simulation setup is described. Subsequently, illustrative results are presented to evaluate performance of the proposed SO-TDMA in comparison with CSMA, PTDMA and PCMAC. Performance is investigated in terms of EC, delay-outage probability, system throughput, fairness, and collision probability.

8.5.1   Simulation Setup and Assumptions

We use MATLAB for simulations of all MAC protocols in the 20 MHz frequency band. We implement the basic CSMA MAC functionality first and test it rigorously to confirm its performance matched with the literature (e.g., [10]) and then implement Algorithms 1 and 2 for the proposed SO-TDMA simulations. The PTDMA and Ideal-PTDMA (or upper bound) are also implemented. For the Ideal-PTDMA, it is assumed each device has updated the knowledge of channel state and queue length information of all devices (i.e., N_a). In particular, the value of $T_{s,d}$ is fixed to $T_{s,d} = T_0$ for CSMA, $T_{s,d} = T_f/N_d$ for PTDMA, and $T_{s,d} = T_f/N_a$ for Ideal-PTDMA. For SO-TDMA, the value of $T_{s,d}$ is optimized as in Algorithm 1.

The frame-length update in PCMAC has been implemented exactly as given in Algorithm 1 of [9]. All PCMAC simulation parameters in the implementation have been kept the same as described in Section IV of [9]. All common MAC simulation parameters like packet size, back-off time-slot size, traffic and channel

Table 8.1 Simulation parameters

Simulation parameter	Value
T_m	50 s
N_d	2–10
B	20 MHz
T	10 μs
I_{th}	$30 \times T$
T_f	$1000 \times T$
T_{min}	$40 \times T$
T_{max}	$T_f - I_{th} = 970 \times T$
T_0	$100T$
T_c	10 ms
D_{max}	50 ms
Average SNR	20 dB
DIFS	$4 \times T$
SIFS	$1 \times T$
P_s	2400 bytes
CW_{min}	16
CW_{max}	1024
α	0.7
ε	0.001
δ	0.5
W_I	$5 \times T$
W_D	0.01

Table 8.2 Transmission rates R_k vs. SNR ranges used in the illustrative results

R_k (Mbps)	SNR (dB) $[\eta_k, \eta_{k+1})$
6	[5, 8)
9	[8, 10)
12	[10, 13)
18	[13, 16)
24	[16, 19)
36	[19, 22)
48	[22, 25)
54	[25,∞)

model characteristics for consistency have been set to the same value assumed in other MAC protocols simulated in this work (listed in Tables 8.1 and 8.2).

Furthermore, we assume a Rayleigh fading channel where the channel power gain has an exponential distribution and is independent and identically distributed for all devices. For the packet arrival, two traffic models are examined in the provided numerical results by feeding devices with (1) Poisson traffic and (2) CBR traffic, where μ_d is the mean arrival bit rate for device d. It should be noted that Poisson packet arrival is selected in most of the numerical results since it would represent the worse case compared with the constant bit rate (CBR) traffic according to Fig. 8.5. However, we present some results to evaluate the performance of SO-TDMA with CBR traffic model as well.

The parameter settings used for all MAC protocols are summarized in Tables 8.1 and 8.2. The values of T, $CW_{\min}$, and $CW_{\max}$ chosen in Table I and the data rates listed in Table II are based on OFDM PHY specifications given in Sections 18.4.4 and 18.1.1 of [17], respectively. Moreover, the values of SIFS and DIFS are chosen as the multiples of the time slot size (i.e., T) for the sake of simplicity in implementation. The relations between T, SIFS and DIFS have been largely kept as in OFDM PHY specifications, although there are minimal differences in SIFS and DIFS from values in [17].

8.5.2 SO-TDMA Convergence

Figure 8.3 illustrates the convergence process of transmission time-slot of different devices using Algorithm 9 for different values of N_a, assuming $W_I = 5$ time-slots and $W_D = 0.05$. It is shown that $T_{s,d}$ in SO-TDMA protocol can converge to the target value (T_f/N_a) without knowledge of N_a. For clear illustration of the algorithm performance, we consider the saturated network scenario where all devices have the non-empty queues. Therefore, the fluctuations in N_a due to traffic have been ignored in this figure. It can be seen that for different values of N_a, SO-TDMA converges to the ideal value in merely *a few pseudo-frames*. However, the convergence time increases with increasing N_a.

Figure 8.4 shows an snapshot of the *resource allocation* procedure in CSMA and SO-TDMA along with successful transmissions (light green) and collisions (dark red) under saturated traffic conditions, assuming $W_I = 5$ time-slots and $W_D = 0.05$. Figure 8.4b confirms how quickly SO-TDMA protocol approaches a network stable state (i.e., Algorithm 9 converges), while it eliminates the collision probability among devices. However, Fig. 8.4a shows that there exists a non-zero collision probability in CSMA, resulting in poor network performance. It can also be observed that the fairness in CSMA is worse in the saturated condition as compared to SO-TDMA. For example, in Fig. 8.4a, the devices 1 and 7 hold the channel for much longer time as compared to devices 3 and 4 that suffer from starvation. In contrast, in SO-TDMA, Fig. 8.4b shows that fairness is improved where all devices transmit sequentially in the network stable state (achieved in 200 ms) and approach the same time-slot length as shown in Fig. 8.3. This approves how the proposed algorithm without additional complexity can result in more stable MTC networks in terms of less collisions and can lead to higher throughput. Now, we can look at the performance metrics and compare these protocols under different traffic scenarios.

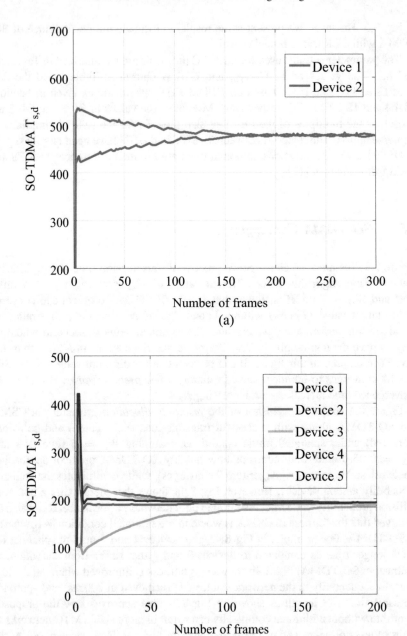

Fig. 8.3 Convergence of T_s for $N_a = 2, 5$ and 10 vs. pseudo frames in SO-TDMA. (a) $N_a = 2$.
(b) $N_a = 5$. (c) $N_a = 10$

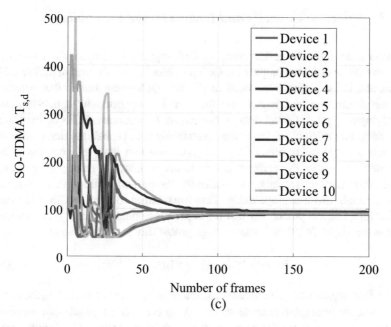

Fig. 8.3 (continued)

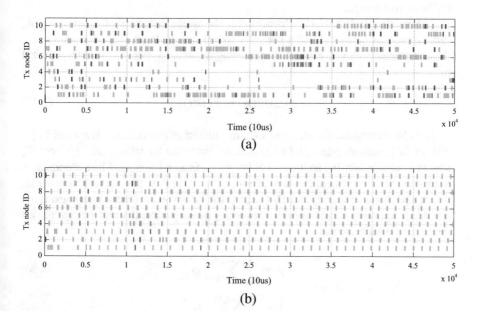

Fig. 8.4 Channel state vs. time ($N_a = 10$, dark red: collision, light green: successful transmission). (**a**) CSMA. (**b**) SO-TDMA

8.5.3 Effective Capacity and Delay Outage Probability

In order to evaluate the performance for QoS support, we empirically calculate EC of a network in order to compare the QoS provisioning capabilities of different MAC protocols. EC was first introduced in [8] as a QoS-aware metric that determines the maximum constant arrival rate that can be supported by a network, while satisfying a target statistical delay requirement. In particular, EC is defined subject to a maximum tolerable *delay-outage* probability (i.e., probability that packet delay exceeds a target delay bound (e.g., D_{max}). Note that the average delay is not a key metric to assess the performance in terms of delay for emerging applications since some packets might be immediately served and some with a large delay with a certain mean average delay. This work rather focuses on the delay outage probability, i.e., the probability that the experienced delay of each packet exceeds a certain threshold. In [8], the delay-outage probability is approximated as

$$\Pr(D(t) \geq D_{max}) \approx \gamma(\mu)e^{-\theta(\mu)D_{max}}, \tag{8.34}$$

where $D(t)$ represents the total delay experienced by a packet, including the queuing delay and the negligible channel service delay (i.e., RF propagation/transmission) at time-slot t, γ denotes the probability of non-empty queue, and θ represents the delay-exponent. It should be noted that both γ and θ are functions of the constant traffic arrival rate μ.

The pair of $\{\gamma(\mu), \theta(\mu)\}$ characterizes the queue behavior of a time-varying service process being offered a constant arrival rate μ. In [8, 18], it is shown that these parameters can be estimated measuring the average delay sensed by a packet as

$$\gamma(\mu)/\theta(\mu) = \mathbb{E}[D(t)]. \tag{8.35}$$

In order to empirically measure EC, an estimation procedure is proposed in [8, 19]. In this method, $\gamma(\mu)$ and $\theta(\mu)$ can be estimated by taking average over N_s samples collected during an interval of length T_m from all devices. More specifically, at the kth sampling epoch, the quantities of $q_k \in \{0, 1\}$ (i.e., the indicator of whether or not a packet is in queue) and D_k (i.e., the total delay experienced by a packet) are recorded. Applying sample means, non-empty queue probability and average packet delay can be estimated as

$$\hat{\gamma} = \frac{1}{N_s} \sum_{k=1}^{N_s} q_k, \tag{8.36}$$

$$\hat{d} = \frac{1}{N_s} \sum_{k=1}^{N_s} D_k. \tag{8.37}$$

Based on (8.35)–(8.37), the estimated delay-exponent becomes

Algorithm 10 EC search algorithm

1: Set $\mu_{\min} = 0$ and $\mu_{\max} = \max R_k$
2: **while** $|\Pr(D(t) \geq D_{\max}) - \varepsilon| \geq (\varepsilon\delta)$ **do**
3:     Set $\mu_d = (\mu_{\min} + \mu_{\max})/2, \forall d = 1, \ldots, N_d$
4:     Run MAC protocol
5:     **if** $\Pr(D(t) \geq D_{\max}) \geq \varepsilon$ **then**
6:         $\mu_{\max} = \mu_d$
7:     **else**
8:         $\mu_{\min} = \mu_d$
9:     **end if**
10: **end while**
11: $\mu_{\text{sys}} = \sum_{d=1}^{N} \mu_d = N\mu_d$

$$\hat{\theta} = \hat{\gamma}/\hat{d}. \tag{8.38}$$

Consequently, employing $\hat{\gamma}$ and $\hat{\theta}$, the delay-outage probability in (8.34) can be approximated as

$$\Pr(D(t) \geq D_{\max}) \approx \hat{\gamma} e^{-\hat{\theta} D_{\max}} \tag{8.39}$$

Given a delay bound $D_{\max}$, based on (8.39), supplying a traffic source with a constant rate μ shall result in a certain delay-outage probability. Thus, by iteratively testing different values of μ, we can reach to our ultimate goal and empirically compute the EC, i.e., the maximum value of μ corresponding to the target delay-outage probability threshold ε. The bisection search method used to find EC is presented in Algorithm 10.

It is worth mentioning that we aim to measure EC for an MTC network, not a single wireless link. We assume that all devices feature homogeneous traffic load. Under this assumption, in Algorithm 10, each device's μ_d is identically varied to find the maximum value of $\mu_{\text{sys}} = N_d\mu_d$ that satisfies the statistical QoS constraint.

Figure 8.5 shows the delay-outage probability in different protocols, measured based on (8.39) versus μ_{sys}. Two traffic models are examined by feeding devices with (1) CBR traffic and (2) Poisson traffic. For the same load (mean traffic rate), Poisson traffic introduces higher delay-outage probability due to its variability/burstiness. For both traffic types, at high load, PTDMA outperforms CSMA since CSMA suffers from a large number of collisions, which are avoided in PTDMA. Figure 8.5 also includes the case of Ideal-PTDMA (in which the time-slot size $T_{s,d}$ is adapted based on N_a) to provide the *lower bound* on the delay-outage probability. The results reveal a need for a *self-organizing* PTDMA protocol that is able to derive this information fast enough in a distributed manner to approach the *lower bound*.

Interestingly, the proposed SO-TDMA achieves a performance very close to the lower bound and outperforms CSMA, PTDMA and PCMAC. Thus, for a fixed acceptable threshold for delay-outage probability (as QoS requirements), much larger effective capacity can be supported in SO-TDMA than CSMA and PTDMA. For example, Fig. 8.5c shows that to maintain a delay-outage probability

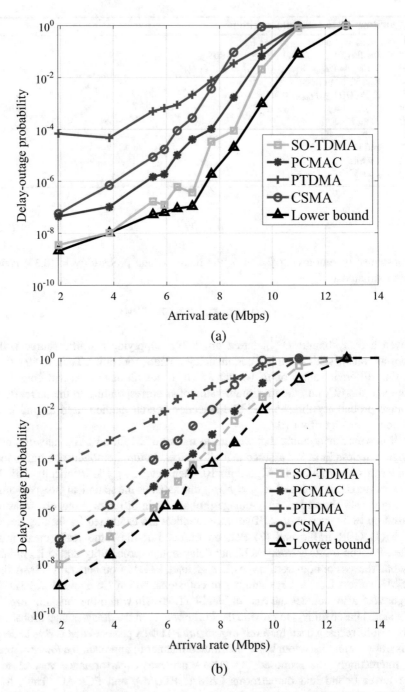

Fig. 8.5 Delay-outage probability vs. arrival rate μ_{sys} (Mbps). (**a**) $N_d = 5$, CBR traffic. (**b**) $N_d = 5$, Poisson traffic. (**c**) $N_d = 10$, Poisson traffic

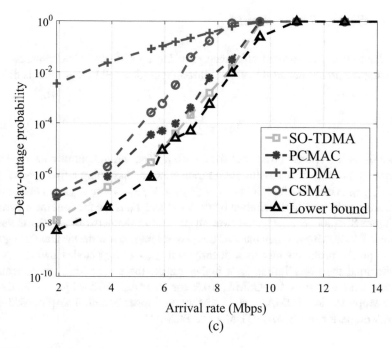

Fig. 8.5 (continued)

not exceeding 10^{-3}, SO-TDMA offers an effective capacity of 7.6 Mbps or 95% of the lower-bound of 8 Mbps, while CSMA achieves 75% and PTDMA has only 15% of the lower-bound of 8 Mbps. This result shows the effectiveness of SO-TDMA in supporting QoS of devices in MTC networks without the need for any additional message passing between devices or a central entity.

The comparison in terms of EC (normalized per number of devices) is shown in Fig. 8.6a, b plotted against the number of devices in the network (N_d) for both CBR and Poisson traffic arrival, respectively. The EC of SO-TDMA protocol approaches the upper bound (i.e., Ideal-PTDMA) and is significantly better than CSMA. Hence, in terms of QoS, the SO-TDMA protocol shows better performance than CSMA, PTDMA and PCMAC for the different values of N_d. For example, it can be observed from Fig. 8.6a, b that the performance gain of SO-TDMA over CSMA is 15–40% in terms of effective capacity. In other words, 15–40% additional multimedia or delay sensitive services can be catered for with SO-TDMA MAC protocol as compared to CSMA with the same QoS requirements.

8.5.4   System Throughput

In order to evaluate spectral efficiency provided by different MAC protocols, we study the maximum achievable system throughput (denoted by S_{sys}) that is defined as

$$S_{\text{sys}} = \sum\nolimits_{d=1}^{N_d} S_d, \tag{8.40}$$

where S_d is the uplink throughput for device d. The throughput of each device is defined as the rate of successful packet transmission. Assuming a fixed packet size (i.e., P_s), S_d is empirically calculated as $(K_{\text{suc},d} P_s)/T_m$ where $K_{\text{suc},d}$ is the number of packets successfully transmitted by device d and T_m is the measurement period.

Figure 8.7 plots S aggregated over all devices in the network as μ_{sys} is varied between 0 to 20 Mbps. When the load (i.e., arrival rate) is low the system throughput S_{sys} in all the protocols increases linearly with μ_{sys} as expected. However, as the system load increases further in a higher range, the system throughput reaches saturation at 9.5 Mbps for CSMA, 10.3 for PCMAC, 10.5 Mbps for PTDMA, 11.25 Mbps for SO-TDMA, and 11.5 Mbps for upper-bound. The proposed SO-TDMA outperforms CSMA, PTDMA and PCMAC.

8.5.5   Fairness

It is known that CSMA suffers from poor fairness performance in high traffic load conditions as studied in [20] and [21]. In case of PTDMA, the resource allocation of each device is equal thus achieving high fairness at pseudo-frame level. However, for SO-TDMA the percentage of channel occupied by an active device in the network is adaptive and optimized independently by each active device as opposed to PTDMA. Therefore, it is very important to look at its fairness performance as compared to other existing protocols.

To illustrate the short-term achievable fairness of the MAC protocols versus N_d, the Jain's index $J(S_1, S_2, \ldots, S_{N_d})$ is computed over non-overlapping 2-s periods and then averaged and plotted in Fig. 8.8. The results indicate that SO-TDMA can keep short-term fairness reasonably high with Jain's fairness index > 0.9 despite its dynamic behavior, confirming the monotonic convergence to fairness of SO-TDMA as stated in Proposition 2. In PTDMA and PCMAC, due to their fixed time-slot durations, every device will have a slot in the frame and apart from some occasional collisions, their fairness is better, close to perfect. On the other hand, SO-TDMA outperforms these protocols in other performance metrics: delay-outage probability (Fig. 8.5), effective capacity (Fig. 8.6), throughput (Fig. 8.7).

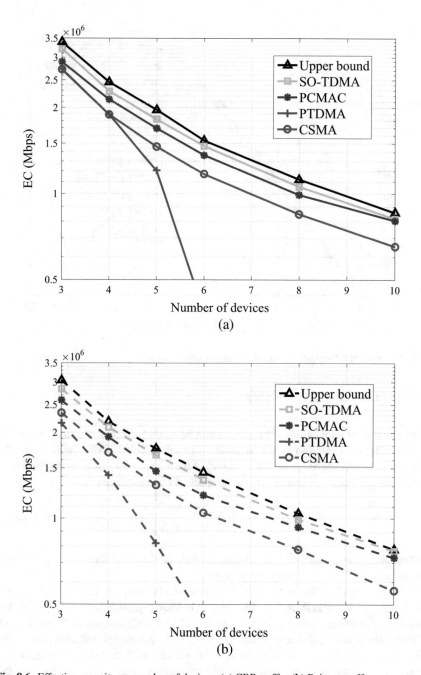

Fig. 8.6 Effective capacity vs. number of devices. (**a**) CBR traffic. (**b**) Poisson traffic

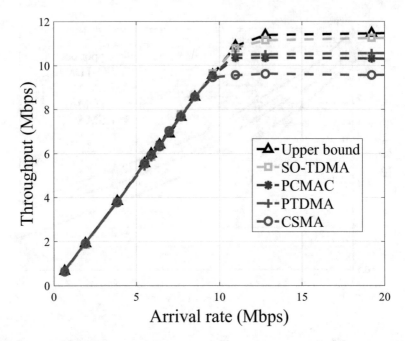

Fig. 8.7 System throughput vs. arrival rate μ_{sys} (Mbps), $N_{\text{d}} = 5$

8.5.6 Collision Probability

One reason for a low system throughput can be a large number of collisions, which are inevitable in random access schemes. Therefore, the collision probability of a contention-based MAC protocol is an important measure to assess its performance. The network collision probability can empirically be calculated as

$$P_{\text{c}} = \frac{1}{N_{\text{d}}} \sum_{d=1}^{N_{\text{d}}} \frac{K_{\text{col},d}}{K_{\text{tot},d}}, \tag{8.41}$$

where $K_{\text{col},d}$ is the number of collided packets of device d and $K_{\text{tot},d}$ is the total number of packets transmitted by device d.

Collision probability in the different protocols is shown in Fig. 8.9 for $N_{\text{d}} = 5$ and $N_{\text{d}} = 10$. PTDMA has the lowest collision probability. SO-TDMA has significantly lower collision probability than CSMA. Compared to PCMAC, SO-TDMA has lower collision probability for $N_{\text{d}} = 5$, and higher collision probability for $N_{\text{d}} = 10$ with high load (i.e., arrival rate > 8 Mbps). But, it should be noted that collision probability of SO-TDMA is still reasonably small for $N_{\text{d}} = 10$ (max 0.3 at worst case). This is mainly because PTDMA and PCMAC operate with a fixed time-slot length, while the time-slot length is dynamic in SO-TDMA.

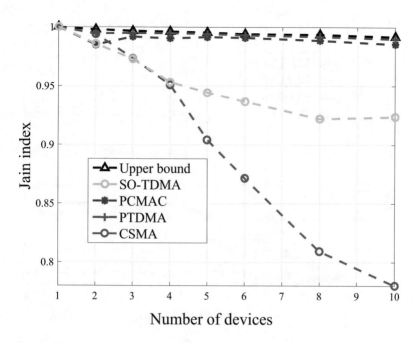

Fig. 8.8 Short-term fairness vs. number of devices

Different from PTDMA with a fixed structure, SO-TDMA needs to learn the frame structure. Thus, increasing the number of devices as well as arrival rates leads to larger convergence times, making it more difficult to approach stability. Thus, a larger number of collisions will happen.

PCMAC is more dynamic than PTDMA as the frame length is adaptive to the number of devices. Compared to SO-TDMA, since PCMAC can stretch out the frame-length relative to the number of devices, each frame will be less congested, and hence, fewer collisions will happen. However, effective capacity would suffer from stretching the frame-length due to the longer delays as shown in Fig. 8.6.

It should be noted that with a slightly greater risk of collisions compared to PTDMA and PCMAC, much better performance in terms of EC has been achieved in SO-TDMA.

8.5.7 Impact of Device Dynamics

To study the impacts of device dynamics on SO-TDMA, Fig. 8.10 illustrates the convergence process of transmission time-slot of different devices using SO-TDMA when more devices join a network. This result shows the transient behavior of SO-TDMA and confirms that SO-TDMA can shortly converge to a stable state after a change in the number of devices. Initially, the network has 8 active devices, and then

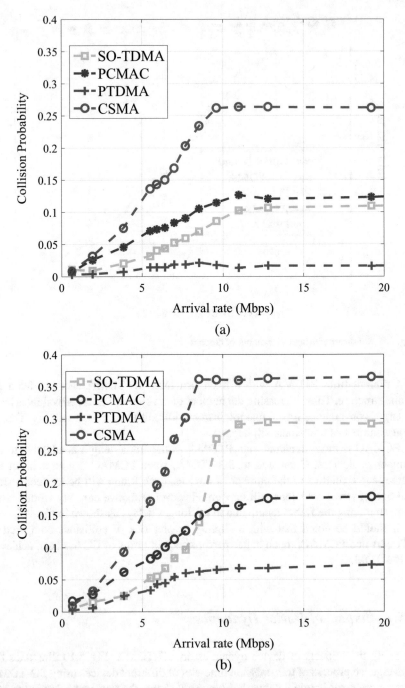

Fig. 8.9 Collision probability vs. arrival rate μ_{sys} (Mbps). (**a**) $N_d = 5$. (**b**) $N_d = 10$

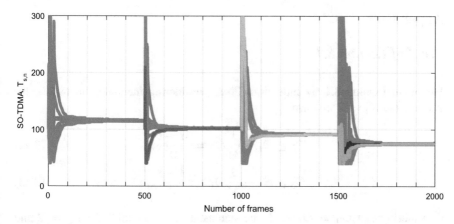

Fig. 8.10 Convergence of T_s vs. pseudo frames in SO-TDMA when new devices join a network of eight devices every 5 s

every 5 s, another device joins the network. The plots for newly joined devices are in colors different from grey. It can be observed that the convergence time of SO-TDMA is short (approximately 1 s). Even though a drastically mobile situation is considered with a very short period of change, the device dynamics could not cause any instability in the network. In the last period, to check how sensitive the stability is to the number of changes, two devices simultaneously join the network. It is shown that SO-TDMA converges to the ideal value of $T_{s,d}$ in a few pseudo frames.

Overall, SO-TDMA outperforms CSMA, PTDMA and PCMAC in terms of EC, delay-outage probability, and system throughput. This is because of the improved and self-organized channel-access by all devices while avoiding collisions. All the results of this section along with aforementioned discussions show how SO-TDMA can improve the spectrum utilization in MTC networks.

8.6 Concluding Remarks

In this chapter, we have proposed the distributed and self-organizing protocol SO-TDMA that utilizes both CSMA and TDMA concepts to reduce the signaling overhead. An AIMD algorithm is proposed for time-slot adaptation process of SO-TDMA, which monotonically converges to fairness. It has been proved that the convergence time required for passing the initial phase in SO-TDMA and moving to the periodic transmission phase is only linearly increasing with respect to the number of devices in the network. Simulation results illustrate the effectiveness of the proposed SO-TDMA in improving channel utilization and effective capacity for MTC networks. The proposed SO-TDMA can approach the performance in the ideal case which needs complete and precise information about the queue length and the channel conditions of all devices.

Appendix

Proof of Lemma 8.1

We prove Lemma 8.1 by induction. First, considering inequality in (8.28), for $N_d = 2$ and $N_d = 3$, we have

$$\Theta_2 \geq \Theta_1 - \frac{1}{M}\Theta_0 = \frac{M-1}{M}. \tag{8.42}$$

$$\Theta_3 \geq \Theta_2 - \frac{2}{M}\Theta_1 \geq \frac{M-3}{M}. \tag{8.43}$$

Considering that $\Theta_0 = \Theta_1 = 1$, the statement in (8.29) holds for $N_d = 2$ and $N_d = 3$. Let assume (8.29) holds for $N_d - 2$ and $N_d - 1$. Then, considering (8.28) for N_d, we have

$$\Theta_{N_d} \geq \Theta_{N_d-1} - \frac{N_d - 1}{M}\Theta_{N_d-2}$$

$$\geq 1 - \frac{(N_d - 1)(N_d - 2)}{2M} - \frac{N_d - 1}{M}\left(1 - \frac{(N_d - 2)(N_d - 3)}{2M}\right)$$

$$\geq 1 - \frac{(N_d - 1)(N_d - 2)}{2M} - \frac{N - 1}{M} = 1 - \frac{N_d(N_d - 1)}{2M}.$$

$\square$

References

1. G. Jakllari, M. Neufeld, R. Ramanathan, A framework for frameless TDMA using slot chains, in *Proceedings of the IEEE International Conference on Mobile Adhoc and Sensor Systems (MASS)* (2012), pp. 56–64
2. X. Chen, Z. Chen, Y. Wu, Leveraging Pseudo-TDMA to a controllable and bandwidth-efficient village area WiFi networks, in *Proceedings of the International Conference on Wireless Communication Network and Mobile Computing* (2008), pp. 1–5
3. I. Tinnirello, P. Gallo, Supporting a Pseudo-TDMA access scheme in mesh wireless networks, in *Wireless Access Flexibility*, vol. 8072 (Springer, Berlin, 2013), pp. 80–92
4. G.S. Paschos, I. Papapanagiotou, S.A. Kotsopoulos, G.K. Karagiannidis, A new MAC protocol with Pseudo-TDMA behavior for supporting quality of service in 802.11 wireless LANs. EURASIP J. Wirel. Commun. Netw. **2006**(1), 065836 (2006)
5. Q. Wei, I. Aad, L. Scalia, J. Widmer, P. Hofmann, L. Loyola, E-MAC: an elastic MAC layer for IEEE 802.11 networks. Wirel. Commun. Mobile Comput. **13**(4), 393–409 (2013)
6. Y. Khan, M. Derakhshani, S. Parsaeefard, T. Le-Ngoc, Self-organizing TDMA MAC protocol for effective capacity improvement in IEEE 802.11 WLANs, in *2015 IEEE Globecom Workshops (GC Wkshps)* (2015), pp. 1–6

7. M. Derakhshani, Y. Khan, D.T. Nguyen, S. Parsaeefard, A.D. Shoaei, T. Le-Ngoc, Self-organizing TDMA: a distributed contention-resolution MAC protocol. IEEE Access (7), 144845–144860 (2019)
8. D. Wu, R. Negi, Effective capacity: a wireless link model for support of quality of service. IEEE Trans. Wirel. Commun. **2**(4), 630–643 (2003)
9. M. Lee, Y. Kim, C.-H. Choi, Period-controlled MAC for high performance in wireless networks. IEEE/ACM Trans. Netw. **19**(4), 1237–1250 (2011)
10. G. Bianchi, Performance analysis of the IEEE 802.11 distributed coordination function. IEEE J. Sel. Areas Commun. **18**(3), 535–547 (2000)
11. D.-M. Chiu, R. Jain, Analysis of the increase and decrease algorithms for congestion avoidance in computer networks. Comput. Netw. ISDN Syst. **17**(1), 1–14 (1989)
12. F.P. Kelly, A.K. Maulloo, D.K. Tan, Rate control for communication networks: shadow prices, proportional fairness and stability. J. Oper. Res. Soc. **49**, 237–252 (1998)
13. D. Loguinov, H. Radha, End-to-end rate-based congestion control: convergence properties and scalability analysis. IEEE/ACM Trans. Netw. **11**(4), 564–577 (2003)
14. J.L. Katz, R.L. Burford, Estimating mean time to absorption for a Markov chain with limited information. Appl. Stochastic Models Data Anal. **4**(4), 217–230 (1988)
15. Y. Huang, W. McColl, Analytical inversion of general tridiagonal matrices. J. Phys. A Math. Gen. **30**(22), 7919 (1997)
16. T. Juhana, R. Hersyandika, Experimental study of TCP unfairness in IEEE802.11-based ad hoc network, in *Proceedings of the International Conference on Telecommunication Systems, Services, and Applications (TSSA)* (2011), pp. 191–194
17. IEEE Std. 802.11, *Wireless LAN Medium Access Control (MAC) and Physical Layer (PHY) Specifications* (IEEE Computer Society, Washington, 2012)
18. B. Mark, G. Ramamurthy, Real-time estimation and dynamic renegotiation of UPC parameters for arbitrary traffic sources in ATM networks. IEEE/ACM Trans. Netw. **6**(6), 811–827 (1998)
19. A. Davy, B. Meskill, J. Domingo-Pascual, An empirical study of effective capacity throughputs in 802.11 wireless networks, in *Proceedings of the IEEE Global Communication Conference (GLOBECOM)* (2012), pp. 1770–1775
20. M.Z. Ying Jian, S. Chen, Achieving MAC-layer fairness in CSMA/CA networks. IEEE/ACM Trans. Netw. **19**(5), 1472–1484 (2011)
21. T. Nandagopal, T.-E. Kim, X. Gao, V. Bharghavan, Achieving MAC layer fairness in wireless packet networks, in *Proceedings of the International Conference on Mobile Computing and Networking* (2000), pp. 87–98

Chapter 9
Conclusions and Future Works

9.1 Summary

In Chap. 3, we have presented an access scheme for a virtualized M2M network aiming to improve the network performance and isolation among network slices. Taking into account the statistical properties of arrival traffic of devices, an MDP is formulated to maximize the network throughput subject to slice reservations. By introducing the policy tree of the MDP, we have presented an optimal access policy. Each device can track this policy tree by carrier sensing and learn its transmission opportunity. As computational complexity of the policy tree grows exponentially with the total number of devices, an efficient heuristic algorithm is proposed based on the MDP formulation, where each device is assigned a deterministic backoff value. Numerical results show that the performance of the proposed heuristic algorithm closely matches to the optimal policy. Moreover, both optimal and heuristic algorithms significantly improve TDMA, CSMA, and DEB in terms of packet delivery ratio and isolation in unsaturated networks.

In Chap. 4, we have presented a reconfigurable access scheme where the partition between demand-free assignment-based and random-based access segments in each frame is adaptive to the network status. In particular, to support a virtualized wireless network consisting of multiple network slices, each having heterogeneous and unsaturated devices, the proposed scheme aims to configure the partition for maximizing network throughput while maintaining the slice reservations. Applying complementary geometric programming and monomial approximations, an iterative algorithm is developed to find the optimal solution. The results show that using this scheme, better performance in terms of throughput and isolation can be achieved in comparison with DQ, which is a recent access scheme proposed for M2M networks.

The partitioning algorithm requires the knowledge of the device traffic statistics. Assuming the absence of such knowledge, in Chap. 5, we have developed a learning algorithm employing Thompson sampling to acquire packet arrival probabilities of

© Springer Nature Switzerland AG 2020

T. Le-Ngoc, A. Dalili Shoaei, *Learning-Based Reconfigurable Multiple Access Schemes for Virtualized MTC Networks*, Wireless Networks, https://doi.org/10.1007/978-3-030-60382-3_9

devices. Furthermore, we have modeled the problem as a thresholding multi-armed bandit and have proposed a threshold-based reconfigurable access scheme, which is proved to achieve the optimal regret bound.

In Chap. 6, we have considered the scenario that devices access the unlicensed band channels through different wireless technologies; LTE and WiFi. To avoid WiFi performance degradation, we have proposed a coordinated structure, in which both networks are controlled by a higher-level network entity. In such a model, LTE devices can transmit in the assigned time-slots, while WiFi devices can compete with each other by using p-persistent CSMA in their exclusive time-share. In an unsaturated network, at each duty cycle, the scheduling for LTE devices and p values for WiFi devices should be efficiently updated by the central controller. The corresponding optimization problem has been formulated and an iterative algorithm has been developed to find the optimal solution. The simulation results reveal the performance gains of the proposed algorithm in preserving the WiFi throughput requirement.

In Chap. 7, we have enhanced the proposed reconfigurable access scheme with NOMA for scenarios of massive machine-type communications. In particular, in this scheme, in each frame, a separate time duration is allocated for each of the NOMA-based, OMA-based, and random access-based segments, where the length of each segment can be optimized. To solve this optimization problem, an iterative algorithm consisting of two sub-problems is proposed. The first sub-problem deals with selecting devices for the NOMA/OMA-based transmissions, while the second one optimizes the parameter of the random access scheme. To show the efficacy of the proposed scheme, the results are compared with the reconfigurable access scheme which does not support NOMA. The results demonstrate that by using a proper device pairing scheme for the NOMA-based transmissions, the proposed reconfigurable scheme achieves better performance when NOMA is adopted.

Finally, in Chap. 8, we have proposed a fully distributed access scheme to reduce the signaling overhead. As in MTC, due to the small size of packets, reducing the signaling overhead makes a significant improvement in channel utilization. In the proposed scheme, devices first use CSMA to access the channel but then converge to TDMA by adopting the proposed learning-based scheme. Furthermore, all devices adjust their time-slot length based on the number of idle time-slots observed in previous frame. Eventually, the time-slot length for all devices converge to the same value ensuring that the scheme is able to provide fairness for all devices.

9.2   Potential Future Studies

In the following, we provide some topics that can be further investigated. In particular, we discuss potential technologies and techniques that can be used to enhance the performance of MTC systems. We also present realistic scenarios that should be considered in these systems.

9.2.1 Correlated Traffic

In this book, we have only investigated scenarios that packets are generated in devices independently. However, there are M2M applications, in which devices transmit packets that are often highly correlated in space and time, e.g., in temperature measurement systems, or in smart grid systems due to the cascading power-grid failures [1, 2]. For these applications, the performance of the proposed reconfigurable access scheme can be improved by taking into account the traffic correlation between devices. To that end, a proper algorithm to detect the traffic correlation between devices should be developed and then the gained information can be leveraged to assign radio resources to devices in more efficient manner. In particular, depending on the application, the correlated data, generated by different devices might be considered as redundant. Thus, for these scenarios, assigning radio resources to each device regardless of its correlation with other devices may result in sub-optimal performance.

Moreover, assuming correlated traffic with unknown parameters the reconfigurable access scheme scheduling problem can be formulated as a global bandit. In global bandits, arms have correlated expected rewards, thus choosing one arm also gives information about the expected rewards of other arms [3]. A Thompson sampling based algorithm can be proposed for these scenarios and the performance of the algorithm can be derived accordingly.

9.2.2 Massive MIMO

As mentioned in Chap. 2, there are different techniques to support massive connectivity including massive MIMO. In this technique, to increase the spectral efficiency, the AP/BS is equipped with antennas much more than the number of devices, so that the channels to the different devices will be quasi-orthogonal. Consequently, using this technique, the number of simultaneous transmissions to the AP/BS can be increased, which is highly desirable in massive MTC scenarios [4, 5]. One interesting subject is to study the performance of the proposed reconfigurable access scheme for scenarios that the AP is enhanced with massive MIMO. In fact, using this technology, the performance of the system can be improved while no additional burden is added to devices. In order to deploy massive MIMO, channel state information needs to be acquired. To that end, orthogonal pilot sequences can be used for both estimating the channel coefficients and then using them for uplink beamforming [6]. However, as there are limited number of pilot sequences, they should be allocated to devices in a dynamic manner.

In fact, by enabling the AP/BS with massive MIMO, in the assignment-based segment of the reconfigurable access scheme, multiple devices are assigned to the same time-slot while each device is allocated a different pilot sequence. The time-slot allocation should be done in a way that rate requirements of devices sharing the same time-slot are met. In the random access segment, the Aloha scheme can be used in which a device randomly chooses a pilot sequence and transmits over a time-slot with probability p. The transmission is considered successful if the pilot sequence is only chosen by one device and the interference from other devices is sufficiently small.

9.2.3 URLLC Use Case

In this book, we have mainly considered machine-type communication scenarios, where the applications were not delay-sensitive. However, there is another category of machine-type communications referred to as ultra reliable and low latency communication (URLLC), where the generated packets should be delivered before a strict deadline with a very high probability [7]. For these time-critical applications, the proposed reconfigurable access scheme should be modified such that these types of requirements can be satisfied. More specifically, in these scenarios, multiple retransmissions should be allowed in order to achieve the required reliable performance. Different from the proposed reconfigurable access scheme, in which each device is allocated at most one time-slot, for these type of devices, multiple time-slots should be allocated to each device in each frame. However, to avoid low spectral efficiency, massive connectivity techniques such as NOMA or massive MIMO should be deployed.

9.2.4 Machine Learning

In order to maximize the performance of access schemes traditionally optimization techniques are used. In particular, the access problem can be formulated as an optimization problem which should be solved at each time-frame by the BS/AP [8, 9]. However, in many M2M applications such as real time applications which are delay sensitive, the time-frame might be considered very small. Therefore, instead of solving the optimization problem, approaches with lower computational complexities are needed to be used. One promising approach is to model the scheduler as a deep neural network, as it requires to perform only limited number of operations to obtain the output. For example, the optimal scheduler proposed for the reconfigurable access scheme can be modeled as a deep neural network (DNN). To do that, the scheduler algorithm should be run for a large amount of different values of input. Then, the input and output should be passed to the DNN for training.

9.2.5 Integrating Massive MIMO, NOMA, Full Duplex

In order to benefit from recent wireless technologies, an access scheme which is able to exploit theses capabilities is needed. In particular, the performance of M2M networks can be maximized by using the combination of NOMA, full duplex and massive MIMO. These technologies are promising solutions to provide massive connectivity and reach high throughput efficiency, low latency, high reliability and high energy efficiency. In this scheme, multiple two-device clusters are formed based on the available number of radio frequency chains (RFCs) at the BS and channel conditions, and NOMA is applied within each cluster [10, 11]. Furthermore, using a reconfigurable access scheme which combines both grant-based and grant-free access schemes, the full duplex capability can help to terminate unsuccessful transmissions happening due to collisions in the grant-free segment or low received SNR at the BS. Moreover, the BS can send a radio frequency energy signal to those devices that are suffering from limited battery life.

References

1. A. Zanella, M. Zorzi, A.F. dos Santos, P. Popovski, N. Pratas, C. Stefanovic, A. Dekorsy, C. Bockelmann, B. Busropan, T.A. Norp, M2M massive wireless access: challenges, research issues, and ways forward, in *IEEE Globecom Workshops (GC Wkshps), Atlanta, GA, USA* (2013)
2. S. Ali, W. Saad, N. Rajatheva, A directed information learning framework for event-driven M2M traffic prediction. IEEE Commun. Lett. **22**(11), 2378–2381 (2018)
3. O. Atan, C. Tekin, M. van der Schaar, Global bandits. IEEE Trans. Neural Netw. Learn. Syst. **29**(12), 5798–5811 (2018)
4. A. Biral, M. Centenaro, A. Zanella, L. Vangelista, M. Zorzi, The challenges of M2M massive access in wireless cellular networks. Digital Commun. Netw. **1**(1), 1–19 (2015)
5. A.-S. Bana, E. De Carvalho, B. Soret, T. Abrão, J.C. Marinello, E.G. Larsson, P. Popovski, Massive MIMO for internet of things (IoT) connectivity. Phys. Commun. **37**, 100859 (2019)
6. E. De Carvalho, E. Björnson, J.H. Sørensen, E.G. Larsson, P. Popovski, Random pilot and data access in massive MIMO for machine-type communications. IEEE Trans. Wireless Commun. **16**(12), 7703–7717 (2017)
7. Z. Li, M.A. Uusitalo, H. Shariatmadari, B. Singh, 5G URLLC: design challenges and system concepts, in *International Symposium on Wireless Communication Systems (ISWCS)* (IEEE, New York, 2018), pp. 1–6
8. H. Sun, X. Chen, Q. Shi, M. Hong, X. Fu, N.D. Sidiropoulos, Learning to optimize: training deep neural networks for interference management. IEEE Trans. Signal Process. **66**(20), 5438–5453 (2018)
9. H. Sun, X. Chen, Q. Shi, M. Hong, X. Fu, N.D. Sidiropoulos, Learning to optimize: training deep neural networks for wireless resource management, in *IEEE International Workshop on Signal Processing Advances in Wireless Communication (SPAWC)* (IEEE, New York, 2017), pp. 1–6
10. M. Zeng, W. Hao, O.A. Dobre, H.V. Poor, Energy-efficient power allocation in uplink mmwave massive MIMO with NOMA. IEEE Trans. Veh. Technol. **68**(3), 3000–3004 (2019)
11. W.A. Al-Hussaibi, F.H. Ali, Efficient user clustering, receive antenna selection, and power allocation algorithms for massive MIMO-NOMA systems. IEEE Access **7**, 31 865–31 882 (2019)

Printed in the United States
by Baker & Taylor Publisher Services